FROM THE

FLIGHT DECK

FROM THE
FLIGHT DECK

PLANE TALK AND SKY SCIENCE

DOUG MORRIS

ECW Press

Copyright © Doug Morris, 2007

Published by ECW PRESS 2120 Queen Street East, Suite 200, Toronto,
Ontario, Canada M4E 1E2

Portions of *From the Flight Deck* were originally published in a different form in
EnRoute, *Wings*, *Weatherwise*, *National Post*, *Toronto Star*, and the *Globe and Mail*.

LIBRARY AND ARCHIVES CANADA CATALOGUING IN PUBLICATION

Morris, Doug
From the flight deck : plane talk and sky science / Doug Morris.

ISBN 13: 978-1-55022-765-9
ISBN 10: 1-55022-765-3

1. Aeronautics, Commercial — Miscellanea. 2. Aeronautics — Miscellanea.
3. Air travel — Miscellanea. I. Title.
TL720.M67 2007   629.13   C2006-9006830-5

Editor: Crissy Boylan
Cover and text design: Tania Craan
Production: Mary Bowness
Front cover photo: Getty Images/James Darell
Author photo: John Leighton
Printed by Marquis

This book is set in Janson

With the publication of *From the Flight Deck* ECW PRESS acknowledges the generous
financial support of the Government of Canada through the Book Publishing
Industry Development Program (BPIDP) for our publishing activities.

Canadä

DISTRIBUTION

CANADA: Jaguar Book Group, 100 Armstrong Ave.,
Georgetown, ON L7G 5S4
UNITED STATES: Independent Publishers Group,
814 North Franklin St., Chicago, IL 60610

Printed and bound in Canada
on 100% post-consumer recycled fiber

ECW PRESS
ecwpress.com

To my three children, Kirsten, Kenzie, and Kirk,
to whom I pass the torch

CONTENTS (CHECKLISTS)

Acknowledgments

For writers, business endeavors, and most things in life, it is said, "find a niche and go after it — and stick to what you know best." Since 9/11, information available to the public about the nuts and bolts of the aviation industry has slowed to a trickle. This book helps to fill that void by enlightening a wide spectrum of readers: from frequent flyers to passengers uneasy with flying to those pining to be pilots. For ones in the "aviation know," my book is for their spouse, family, friends, and that inquisitive neighbor. Below is a list of those who helped make it happen.

EnRoute magazine sits near the top of the "thank you" list. Arjun Basu, now editor-in-chief, saw my writing develop over the last nine years as I wrote for their aviation page. Thank you, Arjun, and your contingent of editors that persevered during my numerous takes on getting it right.

Another of life's lessons — "one good thing leads to another" — compels me to thank the editors and publishers at *Wings*, *Weatherwise*, *National Post*, *Toronto Star*, and the *Globe and Mail* for publishing my articles as I branched out into the writing world.

Had it not been for Jack David, publisher of ECW Press, who saw promise in my articles, this book would still remain wishful thinking. The one who had the daunting task of painstakingly perusing over my articles and coming up with the glue to put it all together was Crissy Boylan. What a task, and yet what a job! Thank you.

I thank recently retired captain Allen Reichert, flight attendant

Mario Chu, and meteorologist Jim Abraham for providing feedback from their perspectives. Thanks also goes out to the many experts in their fields who I relied on for hardcore facts: Warren Lampitt, Joe Forbes, Brian Sherman, Edris Berube, Ben Minicucci, Julie Mailhot, David Clark, Steve Lapczak, and many more.

Pictures are always a great complement to a book of this nature, and I greatly appreciate Brian Losito's submission of some great shots.

A huge impetus for this book was my sister Barb — a confi-dante, mentor, and great friend — who succumbed to cancer at an early age of 49.

Lastly, I thank my encouraging wife, Charlene, and my three supportive children who put up with their father "always being on the computer."

Introduction

You could say the transportation industry is in my blood. My grandfather worked for the Newfoundland railroad, and my father took to the sea. My career choice came easy and early in life: I declared in my grade-nine yearbook that I wanted to be an airline pilot and geared my studies toward the sciences.

Without a mentor to direct me in becoming a pilot, I figured having a university degree was a good stepping-stone. I pursued a Bachelor of Science with a major in physics from Halifax, Nova Scotia's Dalhousie University and began taking flying lessons at age 18. I found some well-paying summer jobs, such as planting trees in British Columbia, to pay for flying lessons and university. Now when descending over the Rockies into the Vancouver airport, if I glance from the flight deck I see tree plantations checkerboarding the terrain below. It's hard to imagine that 25 years ago I was one of the many planters busting my back, throwing thousands of trees into the ground, making money to chase dreams. Tree planting builds a person's character. (If it didn't, it would crack you.) There's a saying among tree planters, "If you survive this job, you will go on to successfully achieve your life goals." From my planting group two others also became airline pilots.

Walking on campus during my final year of university, I realized I would have a commercial pilot license and a degree at age 20. I had everything figured out, or so I thought. Canada's economy was in a recession, a time at which the airline business is one of the first to

In B.C. during my tree planting days

suffer and one of the last to recover. Landing a job as an airline pilot proved tougher than I thought. Flying for the military was my second choice, but I decided to give it a try. So did many others, and life as a military pilot did not pan out for me. With no pilot job in sight, back to university I went, this time to Montreal's McGill University to study meteorology. I hoped this would give me an edge, as well as an opportunity to learn French.

I used to stare at the words on a poster strategically placed over a large crack on the wall of my McGill student-ghetto apartment. Though a little corny, the message is one of life's lessons: "If you have a dream, don't let anything dim it. Keep hoping; keep trying. The sky is the limit." Now I pass these words of encouragement on to my students at the end of my PowerPoint presentations.

Upon graduation, I learned that Environment Canada needed meteorologists, so off I went to Toronto to get certified. The course was tough, with a 60 percent failure rate, but luckily having a commercial pilot license kept me from getting the ax. The instructors saw that I appreciated weather as a user and they respected that I was a pilot.

I forecasted weather in Halifax for over three years. The job entailed around-the-clock shifts writing public, marine, and aviation forecasts for the Maritimes at the Maritimes Weather Center. We here on the east coast experience lots of weather and one way to describe it is "variable" so it was a great place to learn the ropes. I also worked as a civilian forecaster for the military in Halifax. Weather forecasting is both a science and an art. When weather systems didn't behave as predicted, I took it personally — maybe I

My private pilot days

was too conscientious. As the joke goes, it's one of the only jobs out there where you can be wrong so many times and still keep your job. That being said, I believe meteorologists do a great job and they are right more often than you may think — or at least that's my biased opinion.

The pull to become an aviator persisted. I threw in the weather map and began flying for a small charter cargo company. Then Air Atlantic, a commuter airline based in Halifax, took me on — my first real airline job! This milestone also led to my marriage, having told my girlfriend for 10 years, "Let's wait until I get established in the airlines." Then another economic slowdown took hold and my career progression stagnated, until Air Canada's east coast connector, Air Nova, began expanding. I did a "wing walk" to the competition, which proved to be one of my best career moves. Air Canada hired me thereafter. For me, as a Canadian and a pilot, making it to Air Canada was like making it to the NHL.

To borrow the words of a retiring Air Canada captain, "Being a pilot is the best job there is; being an airline pilot is the best pilot job there is; and being an Air Canada pilot is the best airline job there is."

When I joined Air Canada I had logged 8000 flight hours and had five years of postsecondary education. I started as a cruise pilot, a new concept at the time. Transport Canada (the federal government department responsible for Canada's transportation policies) authorized a cruise pilot to occupy either seat, captain or first officer (copilot), during the cruise phase of the flight, but did not permit them to perform takeoffs or landings. This arrangement allowed the other pilot to rest in dedicated crew-rest facilities, therefore helping to ensure a safe flight. The cruise pilot position was jokingly referred to as "yawning for yen" (many routes were to Japan) or "dozing for dollars," but as a supervisor once said, "It's a great way to see the world and take lots of pictures, just like you see in *National Geographic*." He pegged it.

The experience of going from bouncing around in a small turboprop airplane in the Maritimes to seeing the world in a state-of-the-art jumbo jet convinced me the flying public would be interested in my perspective. After years of thinking writing and me were like oil and water, the writing bug bit me soon after Air Canada hired me 11 years ago.

During my crew rests I read books on freelance writing and decided to approach *enRoute* magazine, Air Canada's in-flight publication, about writing aviation or weather articles. My background as a meteorologist carried some weight, and my proposal to write two weather-related articles received the stamp of approval. It just so happened *enRoute* was revamping the magazine and, as they say, timing is everything.

Since then I've written articles for *enRoute*'s aviation page (108 and counting), *Wings* (a Canadian aviation magazine), *Weatherwise* (an American weather publication), *National Post*, *Globe and Mail*, and the *Toronto Star*. People, including pilots, tell me they thoroughly enjoy reading the articles and have learned a lot. I'm often asked how I come up with so many interesting topics for aviation columns. It's easy when you stop to think about how complicated and intricate a modern airliner is. Throw in one of the most talked about topics, the weather, as well as the number of people and departments required to get an airliner and the aviation business in the air, and you'll find a recipe for great ideas.

Did you know 9 out of 10 first conversations begin on the topic of weather?

Through commuting, deadheading, and feedback from my articles, and from teaching weather, I've learned that passengers are inquisitive and many are fascinated by the aviation world. *From the Flight Deck*'s aim is to gather the information and insight from my articles into one comprehensive collection for you, the passenger. So now join me on a long-haul flight from Toronto to Hong Kong, and I'll provide the pilot's perspective on the science, operations, and fun of flying.

CHAPTER ONE

Before-Start Checklist

The Life of a Pilot

Getting into the Business

While this holds true for most goals in life, becoming an airline pilot is no easy task. Sacrifices have to be made. There is no magic way to gain flight experience; you have to log the hours. Gone are the days when pilots were hired at age 18 or 19 with bare-minimum qualifications. Nepotism is out, with hardcore qualifications taking its place. Having said that, the aviation world is rather small; I'm always amazed by who knows whom in this tight-knit community.

A 30-year-plus veteran captain's first airline job interview: "I was 18 years old and had relatives in flight management. I was asked whether I had any job experience. After much thought, and prodding from the interview panel, I remembered I used to have a morning paper route. The interviewers liked the fact that I had work experience. Further prodding revealed I had more qualifications for the job than I thought: I had an afternoon paper route as well."

There are three main routes to becoming an airline pilot. The path I took — taking flying lessons at a flying club — is a popular option. But think about backing it up with postsecondary education; it makes a candidate more appealing to the employer, and it's always good to have something to fall back on, just in case. Flying lessons at a club can be the fastest route, or the slowest, depending on how much cash and time you have on hand. Logging valuable flight time is a challenge, and many new pilots offer sightseeing flights to get an hour's flight in while splitting the cost of renting an airplane. It's funny to think that now I rack up nearly 16 hours in one flight, whereas when starting off in this business you're near to begging for one measly hour.

Training in the military is the second path to an airline pilot job. In the United States, the Air Force, Navy, Army, National Guard, and Marines all train pilots. In Canada, the military pool is much smaller, and even if you enter the service you may end up flying a helicopter. (Something to think about if your goal is the airlines.)

The third route is flight colleges, and there are some great ones throughout Canada and the United States. Many are affiliated with established colleges and universities and provide a well-rounded package. For new recruits, Air Canada, like most airlines, values candidates with postsecondary education, diplomas in aviation, or a military background.

Learning to fly is expensive, and many would-be aviators give it a miss when they discover the poor wages they must endure while trying to gain flight hours. I knew of one company that allowed pilots to fly for free while they collected unemployment insurance as pay. Agreeing to fly for poverty wages just to get their derrières into a cockpit shows how passionate pilots are about their profession.

Aviation is not so much a profession as it is a disease. Because many pilots have scrimped throughout their starting years, thriftiness tends to be a prevailing attribute, or so say flight attendants.

How does a pilot know he has gone too far on the thrifty scale? When another pilot notices.

Seniority

Why do some pilots have three stripes on their tunic sleeves and shirt epaulets while others have four? A first officer (also referred to as copilot) has three; a captain has four. But what's the difference?

Not all flights are created equal. Flying the jumbo Airbus A340 to destinations such as Paris, London, or Tokyo is a better gig than the hectic twice-in-one-day Toronto to La Guardia, New York, route. How are these routes assigned to pilots? For many airline companies, including Air Canada, seniority dictates: the airplane a pilot flies, the base from which the pilot flies, the routes flown, the number of days off, and the specific days worked. Airlines may choose age, experience, or marks achieved in initial training to determine seniority. Nowadays, for pilots hired the same day, the valuable seniority ranking is simply decided by fate, with a number pulled from a hat. This happened during my new-hire class. The same thing happened at another airline I flew for, but luckily I drew number one.

The more senior the pilot, the better the chance of getting his or her preferences for aircraft, bases, destinations (domestic versus overseas) and schedules, vacation, and promotions. If you have seniority number 2000, you won't get promoted until the pilot ahead of you, number 1999, makes his or her choice. To complicate things, there is overall seniority within the airline, seniority among the pilot base, and seniority in the airplane and position held. A pilot holding a very senior number flying as a copilot would start near the bottom of the list if he or she went to captain on the same airplane. (Believe me, lawyers have become rich representing pilots dueling over seniority issues after airlines merge.)

As a cardinal rule, politics, religion, and sex are three topics that should be avoided during polite conversation. Seniority in a merged airline should be the fourth.

Pilots are awarded their monthly schedules by a preference system — the more senior a pilot, the better the schedule. Pilots choose what is most important to them: senior pilots will be home for Christmas and have summer vacation. I've spent Christmas doing laundry in a Toronto hotel readying for my flight to Beijing, so you can guess where I sit in the pecking order. At Air Canada, and for most North American airlines, we bid our preferences using a computer program with seniority as its foundation. Number one gets the cream, whereas the bottom-feeders are invariably on reserve (on call). If you want weekends off, want to avoid red-eye flights, or prefer Paris over Tokyo, the choice is yours — providing you're senior enough.

Generally, pilots fly 75 to 85 hours a month. Unlike pilots who need to be certified on each type of aircraft to fly it, flight attendants are certified on all types.

One adage that circulates among the pilots is, "Stay senior on junior equipment." Others want to fly captain as soon as possible, and still others don't care about the rank and put more emphasis on the size of the airplane. For instance, some pilots prefer to fly as first officer on the "big bus," the Airbus A340, over flying as captain on the "little bus," the Airbus A320. And some prefer to be home with their families every night and might remain a copilot in order to build seniority within the rank and fleet instead of upgrading to captain.

Until very recently, my approach was to gain seniority on the A340 as a first officer, avoiding the temptation of going captain on the "little bus." But considering the major hiccups in the aviation world in the last few years, and after the experience of six-and-a-half years flying internationally, I bit the bullet and I was awarded A320 captain flying domestically. By the time this book hits the bookstores, I hope to have completed all the necessary paperwork and procedures and have my extra stripe as a captain.

Seniority distinguishes the rank of captain and first officer, but that's not all. In a two-pilot flight deck only one can have the final say. That's the captain. Even though the captain and first officer share the duties (meaning they alternate the flying), the final decisions rest with the captain. This holds true for most airlines around the world. Not only do the uniform, seniority, and responsibilities differentiate the two, pay does as well. A first officer makes about 50 to 60 percent of the captain's wage.

As far as pay goes, this too has no simple answer. Pilots start on a flat salary and then, after a few years, enter a formula pay scale with many factors, such as years of service, position held (captain or first officer), aircraft type, night versus day rates, destinations flown (domestic or overseas), and the speed and weight of the airplane (size matters). Other companies simply base salaries on years of service. This strategy discourages pilots from transferring to bigger aircraft simply to earn more, thereby cutting training costs for the airline.

Airline captain: "If only I made as much money as people think I make, had as much time off as my neighbors think I have, and had as much fun on layovers as my wife thinks I have."

Commuting

Surprise is the look most passengers have when I sit next to them in full uniform. I can hear them thinking, "Why is a pilot back here in economy class? Isn't he supposed to be flying the airplane?" I'm a commuter. I chose to live in Halifax, but I fly out of Toronto. There are many pilots who do not live in the city they fly from. Rumor has it some 50 percent of U.S. airline pilots commute or have commuted.

As I write this, yet another full flight closed and pushed back, leaving me at the gate. Luckily there are two more flights that can get me to Toronto in time for work. Living in Halifax but working in Toronto, I'm torn between two cities. Even people committed to an hour's commute must find it hard to fathom driving to an airport, waiting for a flight on a standby-only basis, sitting in an airplane for more than two hours, only to wait in another airport to go to work.

As Maritimers, my wife, children, and I tried to assimilate in Toronto for several years. I adapted well, but the pull to return home for my wife only intensified with time. (It didn't help that her identical twin sister was calling almost every night asking, in a roundabout way, when she was coming home. Perhaps our Toronto residency was doomed from the start, as I tried to separate twins fused by a 40-year bond.)

While I was living in Toronto my company encountered financial turbulence, so I decided to pursue another flying job. This took me to the Middle East, where I picked up a saying from the ex-pats, "Happy wife, happy life." If momma bear is not happy, no one in the family is happy. I also gained perspective — seeing the quality of life there versus what we have in Canada. Toronto is where my job is, but luckily for my family and me, I have the option of commuting. So we packed up the moving truck and scurried back to the east coast, where it's laid back, with a more tranquil way of looking at life.

I'm not alone in my civic indecision; there are many more commuting from various cities across this country. I'm just glad my commute doesn't involve another country, as it does for some.

The agent working the flight is paging my name, and it looks like I'll get on this next flight. As my home life and job tears me between two cities, I know one thing for certain, my wife and family are happy (and that includes my wife's twin sister).

It's been said, "The best thing about being an airline pilot is that you can live wherever you want." It has also been said, "The worst thing about being an airline pilot is that you can live wherever you want."

While some of us travel "incognito" (out of uniform), there's a good chance that the next time you fly you'll run into a pilot commuting to work. Commuting is a way of life for many in the airline industry, not just pilots. Flight attendants, aircraft mechanics, ramp handlers, and ticket agents have all been known to commute.

And don't assume commuters just show up at the gate and grab their assigned business-class seat. Commuters fly standby at a reduced rate (yes, sometimes "standby" means "standing" there and waving "bye" to the plane as it pushes back). If the flights are full commuters wait for the next, always with a plan B. Weather is also a big issue. I'm always checking forecasts to see if anything that may keep me from getting to work on time is lurking on the horizon. Many of us have a commuter pad (think fraternity-style living at the commuter's expense) in their work city. Other commuters realize this will be their way of life to the day they retire, so they opt for better accommodations. As for me, I stay at hotels near the airport.

On one recent flight of mine, Toronto non-stop to Hong Kong, the crew consisted of four commuting pilots: a first officer from St. John's, Newfoundland; a cruise pilot who resides in Calgary, Alberta, but is based in Vancouver; a captain who lives six months in Phoenix and the other six near Lake Muskoka, Ontario; and me, from Halifax. Not only that, but among the Toronto-based flight attendants were two Montrealers, one living in Winnipeg, Manitoba, and one enjoying Florida's climate.

I know of several pilots who commuted most of their careers. Even though it can be challenging, there are always compelling reasons to keep doing it: better schedules, more desirable airplanes and routes, a better or more economical place to live, and a happier spouse.

Deadheading

Commuting isn't the only reason you'll see pilots in uniform taking a seat in the cabin. They may also be deadheaders. Deadheading refers to crew members — flight attendants and pilots — who are being repositioned as part of their flying schedule. Here's one website's definition of a deadhead: "A person traveling on a pass; specifically, an airline crew or crew member in transit." In other words they fly as passengers while on duty. Deadheading is sometimes a result of irregular operations; that is, flights operating at different times than originally scheduled due to factors like maintenance, equipment (aircraft) changes, or inclement weather.

For example, the type of aircraft for a flight may be substituted; a flight was supposed to be on an Airbus A321, but at the last minute a Boeing 767 took its place. The airline has the option and ability to switch aircraft. (This is something to think about when booking a ticket. Some small operators may not be able to find a replacement quickly if the plane goes mechanical, for instance, and your trip may take longer than you bargained for. As the adage goes, "you get what you pay for.") Pilots are only "checked out" on one aircraft type at a time. A Boeing pilot can't just hop in the seat of an Airbus and go flying, nor is it legal for a turboprop pilot to fly a commuter jet. When aircraft are swapped, a pilot might travel as a deadheader on the flight he or she was originally scheduled to fly. Once at the destination, deadheading pilots pick up the next leg of the trip on their aircraft type.

Deadheading is also used for regular repositioning. Some of my deadheading has taken me to Beijing, Tokyo, and London to fly the flight back. On-call (or "reserve" crews) are frequent deadheaders, as they can be repositioned to start a trip from another aircrew base. For example, Vancouver-based pilots may be needed to fly Toronto pairings because of a crew shortage there.

The "flight deck" in an A340 simulator

Keeping up to Speed

It's been a busy afternoon for the captain and the first officer of Flight 101. They've just been advised by air traffic control to go into a holding pattern before landing, and the flight deck is a hub of activity. Checklists are itemized as the two crew members notify flight dispatch, the service director, and the passengers. Soon after, Flight 101 concludes circling in its racetrack holding pattern, touches down for an uneventful landing, and taxies over to the arrival gate. Then the unexpected occurs. The cabin door opens to the outside world, but no bustling airport comes into view. Instead, an instructor steps out of a flight simulator, ahead of the flight crew he has just spent four hours testing.

The flight simulator or "sim," as pilots call it, is a multi-million-dollar computerized marvel. Capable of replicating normal, unusual, and emergency flying conditions, it uses cutting-edge motion, sound, and visual effects to simulate the experience of flying an actual aircraft for training purposes.

For all airline pilots, there's no escaping the flight simulator. Just because a pilot has a license to fly and a good job at an airline doesn't mean the end of exams. Every six months, Air Canada pilots are required to renew their licenses with a flight test. Simulator exercises become routine for all airline pilots regardless of their level of experience. Tested twice a year throughout our careers, we get to know the procedure, which involves a full day of training, complete with a pre-briefing, four hours of simulator training, and a debriefing. On day two, pilots put what they've learned into practice with a three- to four-hour flight test. Many of the simulator's scenarios replicate emergency situations and the tests are never taken lightly. I've flown simulators in Paris, Zurich, Miami, Toronto, and San Diego, and they all have a commonality: it's like walking into a morgue — there are no windows and few smiles.

The simulator may be a way of life for an airline pilot, but for me and many other pilots the accompanying stress is something we would rather avoid. It's about as fun as getting a root canal. (Spouses and kids know to tiptoe around the house during simulator time.) Much of the stress is self-imposed. As one supervisor put it, "Most pilots are A-plus to A caliber, but there's a few Bs and some Cs out there." But how would you feel if your job came with a six-month renewal stipulation? I have, however, met other pilots, believe it or not, who rather enjoy the sim and look forward to their next session.

For years the six-month simulator ride came with a pass or failure. At Air Canada two strikes (failures) in two years meant you were looking for other employment. A failure to pass the upgrade exam to captain either meant automatic dismissal or being frozen as first officer for years. Because some pilot supervisors dished out more of their fair share of failures, their reputations preceded them, with nicknames like the Terminator or Smiling Assassin. Now we "train to standards." The focus has shifted from the threat of failure to the encouragement of learning and professional betterment. That doesn't mean, however, that the pressure to excel doesn't still exist for most pilots, myself included.

Each airplane type has its own simulator, also referred to by pilots as "the box." At Air Canada, these are located in either Toronto's or Vancouver's training facilities. (When things get busy, pilots will use simulators from private companies and other airlines.) These state-of-the-art sims recreate a gamut of scenarios with astounding graphics that would impress even the most avid video gamer. Pilots can be ready for takeoff in Montreal or whisked away to Hong Kong with a push of a button. Thanks to the simulator's realistic capabilities, pilots can take off in next-to-zero

An Embraer simulator from the outside

visibility, fly in day or night settings, or land in a blinding snow-storm. The graphics are so detailed that if I taxi the airplane on a snow-covered ramp and turn the airplane around, I see freshly made tracks behind me. Flashing thunderstorms, vehicles on the runway, low cloud heights — these are just a few of the many things pilots see in the simulator.

The motion experienced in the simulator is also very realistic. A typical simulator sits in a large two-story room and is mounted on powerful hydraulic jacks that create the feeling of acceleration, tur-bulence, turns, climbs, descents, and landings. During the simulation exercises, radios are used for communication between the pilot and the government-approved instructor/examiner. The instructor, who programs the exercises while sitting directly behind the pilot, plays a variety of roles: head flight attendant, air traffic controller, company dispatcher, weather office, and maintenance personnel. Working behind the scenes is the support staff, which maintains control over the high-powered computers that monitor the sim. These simulators, which cost some Can$20-million each, are put to the test almost every day of the year. When they aren't being used for pilot training, they are being fine-tuned by mainte-nance crews for the next session. Simulators typically cost Can$600 to Can$1000 per hour to run — a small cost to preserve safety, which is taken very seriously by pilots. Simulator training means that school is never out for pilots, which is just as well, because, after all, practice makes perfect.

Some airlines, including Air Canada, are now introducing a less stressful approach to keeping pilots up to speed by implementing the Advanced Qualification Program (AQP), whereby proficiency is taught system by system. In other words, a gamut of emergency procedures relating to the hydraulic system is completed before going on to the electrical system, and so on. The frequency for this program is lessened from every six months to every eight months. AQP is also about communication, crew resource management (CRM), and how to work as a team member. Pilots are even marked on their public announcements (PAs).

Apart from tests in the simulator, all pilots must pass an annual medical exam to renew their licenses. Pilots over the age of 40 undergo medicals every six months. Air Canada's own medical department conducts the examinations of the company's more than 3500 pilots, carrying out hearing and eye tests, electrocardiograms, and blood and urine tests.

Once a year, a pilot is paired on an actual flight with a check pilot to ensure that he or she is abiding by Air Canada's standard operating procedures. Government inspectors from Transport Canada also randomly observe simulator flight tests, or actual flights, in order to ensure that pilots are adhering to regulations and have valid personal licenses.

Airline pilots fly only one type of aircraft at a time. If pilots change aircraft, they have to undertake training pertinent to the new aircraft before being cleared for a regular flying schedule. This three-month review includes retraining at ground school, intensive simulator sessions, and a flight test to ensure that they are competent, as well as compliant with Air Canada's mandatory flying practices.

Pilots new to the airline are also closely observed and tested. They must successfully complete a one-year period of probation, culminating in a written exam, review of training records, and an interview. Pilots are issued company and government guides to procedures and regulations, as well as their own aircraft manuals, which they must constantly update. Technical bulletins, airport charts, and maps are updated on a continual basis as well.

All pilots attend an annual one-day ground school that reviews topics such as deicing and the handling of dangerous goods. Pilots then join with flight attendants to practice drills, review coordinated emergency procedures, and attend seminars stressing collective responsibility and reliance on each other.

This elaborate system of testing, documentation, and checks and balances enables pilots to add to their vast amount of knowledge and skills in order to meet the high standards set by their airline. As a result, Air Canada's pilots not only remain proficient, but they perform their job according to some of the highest safety standards you'll ever encounter flying.

Staying Qualified

➢ First day: "take home" test on procedures and systems is marked, briefing, four-hour training session followed by a debriefing.

➢ Second day: flight test.

➢ Some aircraft types now utilize an AQP every eight months.

➢ Medical every six months for pilots over 40, every year for under 40.

➢ A flight check every year on all routes flown.

➢ Annual, recurrent one-day classroom training.

➢ Manuals and charts amended twice per month.

➢ And as one pilot said, "Every day you show up for work you must conduct yourself as if you're being tested."

➢ Number of simulators at Air Canada's training center in Toronto: 8 (3 Airbus A320s, 1 Boeing 777, 2 Boeing 767s, 2 Embraers).

➢ Number of simulators in Vancouver: 4 (1 Airbus A320, 1 Airbus A330/A340, 1 Boeing 767, 1 Boeing 737).

➢ Time required to build a simulator: about 14 months.

➢ Approximate cost of Airbus A340 simulator: Can$22–25 million.

➢ Number of hydraulic jacks (legs) required to simulate sense of motion: 6.

➢ Pressure needed to manipulate the sim by these jacks: 1500 psi (10,342 kilopascals).

➢ Examples of virtual weather conditions created in the sim: zero visibility in fog, snowstorms, thunderstorms, gale-force crosswinds, slippery runways.

> Examples of simulation: sound, sight, and motion, including the banking and turning of the aircraft, accelerations, and the feel of tires as they roll across the bumps and cracks in the runway.

Virtual Weather

It is the requisite dark and stormy night. An intense low-pressure system has invaded the area, thrusting winds directly across the runway at 25 knots (29 mph or 46 km/h). (The unit "knot," a nautical mile per hour, is used by aviators, mariners, and weather people.) Visibility is a mere 1200 feet (366 m). A handful of runway lights can be seen through the fog, providing just enough orientation to take off. The four engines spool up on the jumbo jet, laden with enough fuel to send a small car around the equator 55 times. Runway centerline lights zip by. Frequent rudder inputs are needed to offset the antagonizing crosswind. "Rotate" is heard, but simultaneously the words "Power loss!" bellow into the flight deck as we disappear into the clouds.

I input rudder commands to compensate for the yawing 200-foot (61 m) wingspan caused by the failure of the right outboard engine, and I command the other pilot to secure (shutdown) the engine. We notify air traffic control and ask for the latest weather report. Luckily, the winds have shifted, improving the ceiling to 800 feet (244 m) above ground with horizontal visibility of two miles (3.2 km). We decide to return to the airport. The passengers and service director are briefed and an emergency is declared.

Unfortunately, the precision-instrument approach is unserviceable (see page 128), so we must conduct an approach that does not allow a descent quite as low to the ground. As we descend, we encounter moderate mechanical turbulence due to the gusty surface winds, which causes fluctuating airspeed. The runway finally becomes visible — a welcomed sight!

After landing, the simulator comes to rest on its powerful hydraulic jacks. Four hours of real-life scenarios have ended. My simulator partner and I sigh with relief as we are signed off by the supervisor, validating our pilot licenses for another six months.

Since my first trip in the "box" years ago, simulators have come a long way graphically. The virtual-weather visuals make the experience nearly indistinguishable from reality. These state-of-the-art sims recreate innumerable scenarios.

Ceiling heights can be programmed to induce a go-around (a go-around, or missed approach, is when a pilot aborts a landing). Virtual snow blowing across the runway can trigger disorientating whiteout conditions. Headwinds, tailwinds, crosswinds, and upper winds coupled with slippery runways can all be programmed to increase the stress factor. The simulator instructor has up to 500 malfunctions available to provide in-depth training on the aircraft's abnormal or emergency systems and procedures. One of the few drawbacks is the depiction of cloud bases and tops. They tend to be more defined in the simulator so cloud breakout is quick, unlike the real weather world, where ragged cloud bases are more of a challenge.

Even thunderstorms, with their associated turbulence, lightning, and low-level wind shear (LLWS), can be thrown in (see page 83). It's quite a ride, but limitations exist. Simulated turbulence tends to be less dramatic and more rhythmic. Those who've skirted an actual thunderstorm or heavy convective cloud know convective turbulence is rarely consistent. However, we still fasten our seat belts and keep the shoulder straps in place. Intense, heavy rain showers pounding on the windscreen (sometimes so loud that communication on the radios is nearly impossible) doesn't exist in the video world; nor does hail, the threat of lightning strikes, and the adrenaline levels associated with flying in close proximity to real thunder-bumpers.

Simulating Shear

A phenomenon that really gets the heart pumping in the simulator is LLWS. Pilots practice how to safely get out of it because, if not recognized in time, it has the potential to cause a serious accident. In fact, data from flight data recorders of real accidents can be programmed into the simulator's software. In 1985, the engines to Delta Airline's L1011 on approach into Dallas–Fort Worth, Texas, couldn't spool up fast enough to recover from severe LLWS. This tragic lesson brought the phenomenon of LLWS to the forefront. Today, many airports and new aircraft are better equipped to provide wind-shear warnings. I fly the sophisticated Airbus A340, which has wind-shear detection equipment on board, so when the crisp, loud digitized "Wind shear! Wind shear!" is heard you can rest assured it gets our full and immediate attention.

The only place an airline pilot can be certified for landing in near-zero visibility (called auto-land) is in the simulator. And virtual weather readies us for the real stuff. This procedure requires strict monitoring of instruments, resisting the urge to look outside, and waiting for the "bump" of contact with the ground. It requires great faith in technology — both in the real world and in the sim.

Encounters with volcanic ash, which could cause engine flame-out, are practiced in case we meet the real thing. The acrid smell associated with volcanic ash and other smells are not duplicated in the simulator, so we are briefed on these so we can recognize when the conditions are present.

Simulating Senses

As discussed, the motion in the simulator is very realistic. A room off to the side is jammed with computers that run the machine, sending messages via miles of wires. A pilot's performance is recorded and, if need be, printouts are available to point out where he or she may have wavered. Even the sounds of things like engine thrust, gear and flap extension, and fire bells are recreated by the simulator. Smoke can be introduced, necessitating the use of oxygen masks. Senses of sight, sound, and touch are duplicated, causing the occasional seasoned airline pilot to succumb to motion sickness. And just in case things get a little too hairy, we are told about the simulator's override switch.

Today's virtual weather and other special effects make modern simulators very realistic. So much so that a pilot can obtain a license to fly a particular aircraft without ever touching the controls of the real airplane! In fact, the first time a pilot sees the real airplane may be with a full load of passengers. (Not to worry, the pilot is flying with a line indoctrination captain, a trained instructor, on that first flight.)

CHAPTER TWO

After-Start Checklist

Getting to Know the Aircraft

Before we board our flight to Hong Kong, let's explore what goes into building and maintaining the airplanes we rely on to fly us around the world.

Building an Airplane

The two heavy hitters in aircraft manufacturing are American-based Boeing, and Airbus, headquartered in Toulouse, France. These two companies compete in the large- (jumbo-) and medium-size aircraft market. Because of competition from the new kid on the block, Embraer from Brazil, Canadian-based Bombardier has upped its game and is now more competitive than ever in commuter-size aircraft.

For anyone who's ever wondered about what goes into building an aircraft, the Bombardier Aerospace plant in Downsview, Ontario, is a good place to find out. So that's where I went. You can see the former military base if you look outside the airplane window while approaching Toronto Pearson International Airport from the east. Today the military has pulled out, but there remains an army of engineers, technicians, machinists, and support staff building airplanes at the plant. Building an airplane requires a battalion of engineers, and they make up 23 manufacturing and assembly departments at Bombardier.

Bombardier's Q400 aerofoil under construction

The Downsview plant is one of Canada's oldest aircraft manufacturing sites. Since de Havilland built it in 1929, it has turned out
some of Canada's most historically significant aircraft — from the
speedy Mosquito of World War ii, to the Twin Otter bush plane
and the Beaver. Today it houses an aviation museum and a center
for aerospace medicine, in addition to Bombardier's plant. Perhaps
one day the aircraft currently produced there will also merit their
own chapter in aviation history.

Once inside the plant I heard the sound of rivet guns fusing aircraft pieces. The turboprop Q200, Q300, and Q400 are built
entirely on site, as is the popular Global Express private jet. The
wings for the Lear 45 are also built here and then shipped to
Wichita, Kansas, for final assembly.

With so much development going on under several different
roofs, I decided to focus my attention on the new project, which
everyone is proud of, the Dash 8 Q400 located in Bay 10. Two
brand-new Q400s are turned out each month. It carries double the
passengers of the 200 series at speeds faster than any other turboprop passenger airplane. Air Canada Jazz is the world's
second-largest operator of the Dash 8, and many hope to see the
Q400 complement the fleet. I have over 5000 flight hours on a

The Q400 going through its stress test

Dash 8, and to say things have improved would be an understatement. Bombardier now incorporates a noise and vibration system (NVS) that uses sophisticated dampers, microphones, and computers to cancel out noise and vibration from the propellers. The system was originally designed to make nuclear submarines run more quietly (the Q stands for quiet).

Similar to the manufacturing of automobiles, many airplane parts are built off-site. Mitsubishi in Japan builds the Q400's tail and the rear portion of the main cabin (fuselage) for Bombardier. These components make their way by boat to Vancouver and then make their way across Canada by train to Downsview. The wing is built on-site and is one of the first things seen upon entering the monster building that houses seven Q400s in various stages of construction. The main sections of the fuselage are then mated. The wing is stuffed with fuel tanks, wiring, and flight control cables. Then the 93-foot (28.4 m) wing is attached to the assembled components of the fuselage. The new, more powerful pre-built Pratt & Whitney engines are attached, followed by the landing gear, built by BF Goodrich in Oakville, Ontario.

Functional tests are performed, and I'm told most major systems have dual inspections while on the assembly line. Finally the mass of parts takes the shape of an airplane and it is rolled out to Bay 7 for interiors, preflight checks, its first flight, and then back for its first coat of paint. Bombardier runs the sole runway here and operates it with its own air traffic controllers. A 24-hour emergency-response team acts as a backup during the rigorous tests the airplanes must go through for certification. After further exhaustive tests and checks, the airplane is transferred to the delivery center where its proud owners begin the acceptance procedures.

Along the tour I met the chief pilot. He gave me a pilot-to-pilot briefing of the new gizmos found inside the flight deck and cabin. State-of-the-art "glass" instruments (so-called for the computer screens) have replaced analogue instruments in the flight deck. That new-car smell permeated the air, especially from the leather seats in the cabin. And just like that new car, this particular airplane, destined for Norway, was loaded with options and upgrades.

After seeing the finished product, I was shown a building where a static-and-dynamic tester twists and stresses an actual Q400 to compile much-needed data for engineers. A skeletal system of metal girders with hydraulic motors surrounds the airplane and pulls and tugs on the airframe. To see an airplane subjected to these movements would make any pilot cringe. It's similar to what you see in a furniture store, where seat cushions are repetitively tested thousands of times. The airplane undergoes this stress test for two years, the equivalent of 160,000 cycles (landings and takeoffs), twice the life span of an ordinary airplane.

Bombardier is proud of their products, and they give guided group tours of their facilities so you can witness this process yourself. It's quite something to see how parts morph into an airplane, soon to join the ranks of others flying overhead, en route to airports around the world.

Paint by Numbers

When you think about the most important parts of an airplane, the wings may come to mind, or perhaps the instruments in the cockpit. But what about the paint? Probably not. On a recent visit to Air Canada's paint shop, I was amazed to see what an elaborate effort goes into painting the fleet.

In a massive Toronto hangar capable of accommodating the

The old layer of paint is sanded off . . .

smallest Dash or the mammoth Airbus A340 jumbo jet, a staff of 29 works to ensure that all of Air Canada's fleet looks its best. Each aircraft receives a fresh coat of paint every five to six years.

The preparation and painting of an airplane takes six days. Step one is a soap-and-water wash to remove grease, grime, and exhaust residue. They use ordinary water hoses, as pressurized nozzles could damage sensitive probes and antennas. The airplane is now ready for the second step: stripping it to its bare metal. Sensitive instruments, vents, and openings are all taped and covered. No parts are removed except for the windshield wipers. If you've ever done any painting, you know the preparation often takes longer than the actual painting. Painting an airplane is no exception. After the aluminum is exposed, a thorough sanding is done to ensure that the entire surface is free of old paint. Another wash, then the airplane is remasked with tape. Next is an acid wash, to etch the surface for bonding, followed by an application of alodine, a chemical that protects the skin of the airplane. One last wash, then it is remasked for painting with a primer. The spraying of the final coat starts at the tail. Finally, on day six, the finishing touches: the logos, aircraft registration, and markings are stenciled on.

To be certified for use, airplane paint must undergo exhaustive tests. It has to be extremely durable, adhere well to the skin of the

... before the new layer is sprayed on

craft, and keep its shine for five years or more. It is also astronomically expensive. White paint costs approximately Can$450 for a kit that comprises two gallons of paint, one of activator and one of thinner.

When painting a house, the weight of the paint isn't a factor, but it is an important consideration for an airplane. A complete paint job for a Boeing 767 weighs 1482 pounds (672 kg) and is measured within grams to ensure it doesn't upset the aerodynamics of the aircraft.

You may wonder why airlines like Air Canada don't leave most of the fuselage its original bright silvery aluminum. Some airlines do, to save on weight and consequently on fuel costs. But Air Canada feels that, all things considered, painting their airplanes is actually more cost-efficient: the paint protects the skin, and a bare aluminum fuselage requires frequent polishing to keep its shine. An added benefit is that Air Canada's signature red maple leaf makes the aircraft easy to identify no matter what airport in the world it visits.

Brushing Up
- Weight of paint used on the Boeing 767: 1482 lb. (672 kg).
- Total thickness of paint application: 6–7 mils (1 mil or military unit = 0.1016 mm).

- Approximate square footage covered by 1 paint kit (2 gallons of paint, 1 gallon of activator, 1 gallon of thinner): 300 (28 m^2).
- Number of coats applied to each aircraft: 4 (1 coat of primer, 2 of mica or color, 1 clear).
- Name of new blue color exclusive to Air Canada: AC Ice Blue.
- Length of time before an Air Canada aircraft is repainted: 5 or 6 years.
- Last details to be added to the aircraft: logo, aircraft registration, and other markings.
- Location of Air Canada's only paint shop: Toronto.
- Number of painters at Toronto shop: 29.
- Starting point for paint job: aircraft tail.
- Total hours of work required to paint a small aircraft like the Dash 8-100: 850–1200.
- Total hours of work required to paint a larger aircraft like the Boeing 767: 1800–2500.
- Number of gallons of paint used on the Boeing 767: 177 (670 L).

Taking Wing

The Airbus A340-300 I fly has a wingspan of 197 feet and 10 inches (60 m); some runways are only 150 feet (46 m) wide, so the wingtips actually hang over the edges. The larger A340-500, which is also on my license, has a wingspan of 208 feet and 2 inches (63 m), so its wingtips even spread over the wider 200-foot (61 m) runways. These technological marvels have precise bends, cambers, and wingtips that outperform any wing in commercial aviation history.

In centuries past, aspiring aviators experimented with wing design in the hope of one day discovering the secret of flight. Many mimicked birds' wings on the flying machines they invented, but the right combination of aerodynamics, shape, and weight eluded them. Then, in the 18th century, Swiss scientist Daniel Bernoulli came up with a physical principle that would prove to be a fundamental aspect of aircraft-wing design. He discovered that as the velocity of a fluid (or gas) increases, its pressure decreases. Bernoulli's principle can be seen in action when you blow air over a sheet of paper; the stream of air will lower the pressure above the sheet. The natural tendency is for the greater pressure below the paper to push toward the low-pressure zone created above, and if the paper is light enough, this imbalance will cause it to lift.

Nearly two centuries later, the Wright brothers believed that this lift effect was the key to unlocking the mystery of flight. "Whether [the source of lift] be a rising current or something else, it is as well able to support a flying machine as a bird, if man once learns the art of utilizing it," said Wilbur Wright in 1901. Two years later, he and his brother, Orville, made history with the first-ever manned airplane flight.

Achieving lift is a key concern in the design of an airplane's wings, but many other factors also come into play. To keep the weight as low as possible while maintaining maximum strength, wings are made of composite materials and aluminum. Their shape and size varies according to the airplane's needs. Nothing short of a technological marvel, an airliner's wings are precisely angled and curved from the leading to the trailing edges, and from the root to the tips — designed to handle the many intricate laws of aerodynamics in play.

A jet fighter has short, thin, swept-back wings to allow for higher speeds and sharper turns, whereas an airliner's long, thick wings provide more lift and stability.

Engineers have also designed moveable parts that change the shape of the wing to accommodate various tasks. For example, flaps at the back of the wing increase lift for takeoff and slow the aircraft before landing. They are stowed during cruising and are controlled by hydraulic motors. You may have seen them extend in increments as the pilot configures the airplane for landing.

Most airliners also use slats, which are moveable extensions of the wing's leading edge that produce extra lift. On top of the back part of the wing many airplanes have rectangular boards that extend upward, which are called air brakes. They are used to slow the airplane down or to increase the rate of descent. Toward the wing's ends and back edges are moveable rectangular panels called ailerons, which bank (or steer) the airplane. If the left aileron goes up, the one on the right wing automatically goes down, and the aircraft rolls to the left.

Next to the ailerons are spoilers, which look very similar to ailerons and also aid in banking the aircraft. Upon landing, they double as ground spoilers, extending up from the wing to slow the airplane down — much like a deployed chute behind a drag racer. Nowadays, many wings are built with winglets that extend upward at the wing's ends. Winglets reduce the turbulence in the airflow over the wingtips and can increase a wing's efficiency by three to five percent.

Aside from lifting the airplane, wings provide storage space for literally tons of fuel, as well as vast quantities of cables, wires, fuel pumps, hydraulic motors, and fuel sensors. Many aircraft also use the wings to mount the engines and store the landing gear. Inside a wing's panels are metal spars and ribs. Tubes run along the inside leading edge, carrying warm air from the engines to act as a deicer, and along the back edge of the wing are static wicks, which are metal rods or wires designed to release any buildup of static electricity.

The wing of a modern airliner is truly a wonder of physics when you consider the intense amount of lift required to support a 372-metric-ton (410-ton) aircraft speeding through the sky at 80 percent of the speed of sound. Sometimes, if the air is very moist, the intense low pressure above the wings can cause the air to quickly condense, forming a cloud directly over the wing — yet another aspect of the physics of flight you can watch out for from your window seat.

- -

Winging It
➤ Technical name for a wing: airfoil.
➤ Longest wingspan for an airliner: 212 ft. 7 in. (65 m) – Boeing 777.
➤ Soon-to-be the longest wingspan for an airliner: 261 ft. 10 in. (80 m) – Airbus 380 (when it makes its debut).
➤ Shortest wingspan in Air Canada Jazz's jet fleet: 69 ft. 6 in. (21 m) – Canadair CRJ.
➤ Items found inside an airliner's wings: fuel tanks, cables, wires, heating pipes.
➤ Four forces acting on a wing: thrust, drag, lift, weight.
➤ Scientist who postulated how lift is produced by a wing: Daniel Bernoulli (Bernoulli's principle).
➤ Name of device that helps facilitate lift: flap.
➤ Device on upper surface that hinges upward to facilitate descent: speed brake.
➤ Name for curvature of wing: camber.
➤ Auxiliary airfoil located on leading (front) edge of wing: slat.
➤ Moveable surface located near trailing (rear) edge of wing to control lateral roll balance: aileron.
➤ Vertical device located on wing tip that helps reduce drag: winglet.

➤ Air Canada aircraft without winglets: Boeing 767 and 777.

➤ Wires located on trailing edge of wing to discharge static electricity: static wicks.

Next time you get a window seat overlooking the wing, try to identify some of its components. After marveling at the wing's beauty and complexity, take a closer look at what you're looking through: the airplane window.

Window on the World

Not too many people stop to think about the window itself when there's so much to look at from 35,000 feet (10,668 m) above sea level.

This porthole to the outside is usually round or oval, but it's sometimes square and varies in size. You'll notice helicopters and smaller airplanes, which are not pressurized, have large windows, similar to the size of car windows. But the higher an airplane can soar the smaller its windows must be. The small size of commercial aircraft windows is necessary to withstand the forces of a pressurized cabin. This is why, for example, the windows of the Concorde — which operated at a very high altitude and cabin pressure — are so tiny.

Ever wonder what airplane windows are made of? Each window is actually comprised of two panes; both the inner and outer pane are made of acrylic. Sometimes when looking out the window you may notice it appears to be crystalline or has a milky glaze to it. This is a normal phenomenon called crazing, which is caused by water being slowly excreted out of the window at high altitudes because of inside–outside pressure differences. Windows also lose their gleam due to atmospheric conditions, such as abrasive volcanic ash. The eruption of Mount St. Helens was particularly hard on aircraft.

Like anything else on an airplane, windows are scrutinized for wear and tear, and any scratches are measured and monitored. There are rigorous specifications for windows: when a pane nears the limit for minimum thickness, it will be replaced. When it is time for one of Air Canada's windows to be repaired, it is either refurbished or replaced in the window shop located in Montreal. Refurbishing windows requires the surfaces to be sanded with

The view through the double-paned passenger window

slurry (a combination of water and a powdery substance). This ensures your view is a good one.

The flight deck (my office) has completely different windows. They are larger and made of glass. Some can open and are, consequently, much more expensive. Just one half of the front window on the A340 is priced at Can$45,000. Next time you get a chip in your car window just be thankful it's not an airplane window. The two front windows, depending on the manufacturer, usually consist of three glass panes with electronic coils sandwiched between them that heat the panes. All flight deck windshields must meet stringent requirements in order to withstand incidents such as the impact of a bird. Flight deck windows must be certified to handle bird strikes at various airspeeds. One site in Canada where this is tested is at the National Research Council (NRC) in Ottawa, Ontario. An actual bird is fired through a cannon-type launcher into the flight deck window.

Having windows makes the cabin a much more pleasurable environment. At a cruising altitude of 39,000 feet (11,887 m) on a clear day you can see some 240 miles (386 km), giving new meaning to a room with a view.

Boeing's futuristic 787 Dreamliner, slated to join Air Canada's fleet in 2010, will have larger-than-normal passenger windows with an "electrochromic" feature allowing passengers to electronically shade their windows, replacing the plastic sliding screens.

Some facts to ponder: If you've ever wondered how far you can see out the window as a function of altitude this is what to do (you'll need a calculator). Take the square root of your altitude in feet and multiply by 1.23 for a value in miles. For metric units it's the square root of your altitude in meters multiplied by 3.5, which gives a value in kilometers.

Two words to describe the modern-day jet engine are efficient and reliable. The engines for the Airbus A340-500 I fly are made by Rolls-Royce. Not only do they supply a hefty 53,000 pounds of thrust, they also provide electrical power for the entire airplane, drive pumps to pressurize the hydraulic system, supply air to pressurize the cabin, deliver warm air to deice the leading edges of the wings, and more.

Our flight to Hong Kong will involve stepping to higher altitudes as the fuel load lessens and the aircraft lightens. Generally, the higher the better because jet engines are more efficient the higher they go. They pack a punch and they have some interesting characteristics.

Forward Force

An Embraer engine at rest

The advent of the jet engine during World War II ushered in a new era in aviation. Airplanes could now fly higher and faster. Recent advances in engine technology are helping today's jets fly more quietly, efficiently, and reliably than ever before.

Just how does a jet engine work? Newton's Third Law states that for every action there is an equal and opposite reaction. Simply stated, pushing air rearward moves the aircraft forward, in the same way that when you release the end of an inflated balloon the escaping air propels it across the room. A jet engine takes in air and compresses it. Combustion then takes place, and the engine expels the air back into the atmosphere at a velocity higher than the air outside. One of my aviation books describes this as "sucking, squeezing, banging, and blowing." This process creates the thrust necessary to give the airplane forward momentum.

Newer airliners, like the Airbus A340-500, have four high-bypass jet engines with intake fans over eight feet (2.4 m) in diameter. High-bypass engines are known for their fuel efficiency and significantly quieter ride. These engines are at their most fuel efficient at higher altitudes where the air is thinner. It's the combination of airframe, engine, altitude, and speed that provides fuel efficiency at high altitudes. The engine might burn less fuel at high altitudes, but the thrust is also much less. At cruise altitude an engine produces much less thrust, but the thinner air allows the aircraft to fly faster. This is the primary reason for flying at over 35,000 feet (10,668 m). Since winds generally increase with height, they become a significant factor at such high altitudes, so flight dispatchers are constantly juggling all the variables to find the ideal altitude for a given day.

Airliners with larger engines need an additional air source, which comes from either a ground power cart or an auxiliary power unit (APU), a small jet engine located in the tail. The APU also supplies electrical power and air conditioning to the aircraft while it is on the ground.

Pilots control jet engines from the flight deck by using thrust levers. Air Canada's entire Airbus fleet has an auto-thrust system similar to a car's cruise control. When power is needed to climb higher, the onboard computers send a signal to the engine to produce more power without the pilot having to touch the thrust levers. On landing, reversing the thrust helps the airplane decelerate by deflecting air forward. That's the rumbling noise you hear seconds after the plane touches down.

Most of the engines in Air Canada's Airbus fleet are manufactured by CFM, one of the largest jet engine makers in the world. Amazingly, every five seconds a CFM-powered airplane takes off somewhere in the world. CFM combines engineering expertise with the services of two major aircraft engine manufacturers: Snecma of France and GE-Aviation in the U.S. Other top engine manufacturers are Rolls-Royce, Pratt & Whitney, and General Electric, which give the boost to the remainder of Air Canada's lineup.

Jet Pack

➤ The only place Air Canada overhauls jet engines (internally): Montreal. (They are removed from the airplane for an overhaul. Most scheduled maintenance on jet engines is done while they're still attached.)

➤ The number of jet engines Air Canada owns: zero. (They're all leased.)

➤ The percentage of jet engines Air Canada services in-house: 45%. (Of the 9 different engine types 4 are serviced in-house and 5 are outsourced.)

➤ The percentage of jet engines built by the 4 major aircraft builders: zero.

➤ The 3 main engine manufacturers: Pratt & Whitney, Rolls-Royce, GE (General Electric).

➤ Typical number of hours flown before a major overhaul is performed: 15,000 to 20,000.

➤ The amount of takeoffs and landings this translates into: 5000.

➤ The number of days it takes to completely overhaul a B767 engine: 90.

➤ The device used to inspect tight or difficult spots of the engine: boro-scope.

➤ The number one engine parameter scrutinized by maintenance: EGT (exhaust gas temperature).

➤ Most powerful jet engine found on an aircraft: the new Boeing 777-300 ER engine with a thrust setting of 115,300 lb.

Fill 'er Up

You now understand how the wings work and know their many moveable parts. But the inside of the wing also plays another major role for an aircraft: fuel storage. And for a jumbo jet that translates into tons and tons of aviation fuel. The wings are subdivided into several fuel tanks with fuel pumps, backup pumps, and accurate fuel gauges. Some airplanes also store fuel in the center portion of the belly and tail.

When the little refuel light comes on in the cockpit of our Airbus A340-500 series, it signals a vitally important and surprisingly complex step in the flight preparations. During the nearly two hours that this light is on, 140 tons (127 metric tons) of jet fuel will be delivered to top up the tanks to 154 tons (140 metric tons) for a long-haul flight from Toronto to Hong Kong. The amount of fuel is carefully calculated and allocated.

Preparations for our flight actually begin hours before that little light goes on. To calculate the ideal fuel load for the trip, a dispatcher uses a sophisticated flight-planning program that incorporates wind and weather forecasts, the most efficient flying altitudes, available routes, and other factors. According to our flight plan, the A340 will fly for six hours over northern Canada, consuming 8.8 tons (8 metric tons) of fuel per hour.

Throughout the flight, fuel will be moved backward and forward to adjust the airplane's center of gravity, which will reduce aerodynamic drag and conserve fuel. Near the North Pole, the aircraft will have burned off enough fuel so it will be lighter and its optimum altitude will therefore be higher. For greater efficiency, the lighter aircraft then climbs to a higher altitude, further cutting the drag so the aircraft will need less thrust to fly. As a result, the fuel-burn rate drops.

The fuel load must also be sufficient to get us to an alternate airport in case of bad weather. On this flight to Hong Kong, the alternate is Macao, China. This flight also carries an extra four tons of emergency fuel designated as "fuel in tanks," not meant to be used in normal circumstances. It is the fuel necessary to carry the aircraft for an extra 30 minutes beyond the most cautious contingency plan. After taking all the factors into account, the dispatcher then calculates the fuel load to within 100 kilograms (220 lb.) and the flight time to within the minute.

When the refuel light on the flight deck goes out, the fueler enters the flight deck with the fuel slip. My job is to ensure that all the readings tally. The distribution of that fuel is also crucial. It is stored in multiple tanks inside the main wings, and a few additional tons are stowed in the tail. I signal to the fueler that the fuel load is correct and perfectly balanced.

Fuel is the greatest single expenditure for most airlines, aside from salaries, so saving fuel is always a consideration — especially in these days of skyrocketing prices. Throughout our trip the flight deck crew and dispatchers keep a watchful eye on the routing and flight level in order to economize fuel. Our routing takes us north into Russian airspace to avoid strong 155-mile-per-hour (250 km/h) headwinds. In Hong Kong the weather is good, as predicted, and fuel consumption is spot-on.

On landing, the captain asks for two of the jet's four engines to be shut down during the taxi, again saving precious fuel. After 16 hours in the air, the airplane noses into the gate, then we set the park brake and shut down the remaining two engines. When the captain and I finish our checks in the flight deck, a fresh crew arrives for the return flight. Just as we begin briefing them, the refuel light goes on and the process begins again.

Lots of Gas

➤ Difference between jet fuel and gasoline: jet fuel requires less refining.

➤ Fuel requirements are determined by: flight dispatchers (and confirmed by the flight crew).

➤ Fuel-burn rates for aircraft's cruise altitudes: A340-500 – 8000 kg/h (17,637 lb./h); A340-300 – 6500 kg/h (14,330 lb./h); RJ Challenger – 1300 kg/h (2866 lb./h).

➤ Number of metric tons of fuel required for a 14-hour Toronto-to-Shanghai, China, flight on an A340: 105 (115 tons).

➤ Number of metric tons required for a Toronto-to-Montreal flight on the RJ Challenger: 2.3 (2.5 tons).

➤ Factors that may affect fuel consumption: headwinds, tailwinds, outside air temperature, altitude, aircraft weight.

➤ Color of jet fuel: straw-colored or clear.

➤ Weight of 1 L of jet fuel at 15°C (59°F): 0.84 kg (1.9 lb.).

➤ Name of flight-planning system that determines Air Canada's fuel calculations: AFPAC (Automatic Flight Planning Air Canada).

➤ An AFPAC flight plan calculates required fuel burns to within: 100 kg (220 lb.).

➤ During taxi a Boeing 777 burns: 1633 kg/h (3600 lb/h).

➤ Amount of jet fuel purchased last year by Air Canada: 3.8-billion L.

➤ Fuel instruments in flight deck: fuel quantity, fuel used, fuel flow, fuel temperature, fuel pressure.

➤ Temperature in jet engine combustion chamber at which fuel ignites: 700°C (1292°F).

To give you some idea as to the magnitude of the A340, the fuel it requires, and power bursting from its four engines, let's compare the two vehicles I drive.

As I backed out of the driveway on my way to the airport, my neighbor gave me a thumbs-up for my new license plate frame: "My other vehicle is an Airbus A340." My dependable '99 Honda Civic is no match for the Airbus A340-500, currently the largest Canadian-registered passenger aircraft.

Fuel

The fuel requirement for the nearly 16-hour flight to Hong Kong is 178,000 liters (47,000 gal.). It will take about two hours to fill the wing, belly, and tail fuel tanks with enough fuel for 3,600 Civics. Here's another way to look at it: the fuel needed to complete this flight could send my Honda Civic around the equator 80 times. My Civic fuel consumption pales in comparison to the guzzle rate of more than 10,000 liters per hour (2640 gal./h) of this four-engine jumbo. But bear in mind that flying is still deemed the most economical way to travel.

Wheels

Posing with the A340 left main landing gear

It takes 14 wheels to support the weight of my airliner, with tires made by Goodyear. Each main wheel (tire and rim) weighs 530 pounds (240 kg), with a price tag of Can$2100 for one tire. Each tire has its own individual file and can be retreaded, but there are strict guidelines on wear. For racing drivers, laying some rubber is part of their job, but for pilots the challenge is a smooth landing that leaves as little rubber as possible. Unlike a car tire, an aircraft tire is filled with nitrogen. Air contains moisture and would freeze at high altitudes, with the average temperature at −57°C (−71°F). Nitrogen doesn't form a liquid until −173°C (−279°F) and contains little to no moisture.

As well, landings subject tires to tremendous forces, as they must accelerate very quickly. They may heat up quickly, and with air containing approximately 20 percent oxygen, a fire threat would loom. This is virtually eliminated by using nitrogen. This unreactive gas also prolongs tire life by not allowing oxidation and rust formation inside the wheel.

An Airbus A340-500 tips the scales at 372 metric tons (410 tons) — about 372 Civics. A rejected takeoff therefore requires the braking power needed to stop the momentum of 372 cars moving at speeds nearing 190 miles per hour (300 km/h). Carbon brakes are preferred because they are more durable, but they're also much pricier. Two pedals in front of each pilot control the brakes, but they differ from those on a car by having a left and right set of brakes. The A340-500 also has reverse thrust to slow down and spoilers that deploy, acting much the same way as a parachute in high-performance dragsters.

Power

It takes the "big bus" 56 seconds to reach airborne rotation speed. These speeds vary and are based on weight, wind, temperature, runway conditions, and so on. Today's rotation will occur at 170 knots (196 mph or 315 km/h), requiring more than 6000 feet (2 km) of pavement or concrete, with the end of the runway approaching fast. The maximum speed of 190 knots (219 mph or 350 km/h) for an F1 racer is equivalent to the approach speed of an airliner a few miles from landing

The 110 hp under my Civic's hood is outranked by 53,000 pounds of thrust from each of the Airbus' four Rolls-Royce engines. There's no easy conversion of horsepower to pounds of thrust, but I can assure you that when the thrust levers are advanced to takeoff power, it brings a smile to most pilots' faces.

Airplane Tails

Some of the most exuberant aviation enthusiasts are airplane watchers. Robert Milton, chairman, president, and chief executive officer of Air Canada's parent holding company, admits that he has been watching airplanes since age 10. I frequently see airplane gazers standing near the outer fences of airport perimeters or sitting in their cars, radio scanners tuned in, listening to aviation

Keep your eye on the tail to spot an aircraft's FIN

lingo. Recently I met a couple from London, England, while on the passenger-transfer bus. The couple's trip to Toronto would mainly be spent watching airplanes. In fact, even while I was talking to them the husband stopped the conversation to have his wife write down newfound airplane registrations. Devoted aviation buffs like this British couple jot down registration numbers and pull out their cameras as airplanes pass overhead. More often than not, what they are looking at is one of the craft's most fascinating features: the tail.

On the fuselage portion of the tail is a marker to indicate the airplane's country of origin. This registration number begins with a code letter. In Canada, registrations start with C, and in the United States it's N. Most airlines also include a flight identification number, or FIN, on the tail to differentiate the aircraft in their fleet. For example, Air Canada's Airbus A320 and A319 fleet has numbers in the 200s, and their Boeing 767s have numbers in the 600s. Air Canada's main fleet has a total of 300 FINS. Watchers may occasionally catch sight of FIN 264, an Airbus A319 painted in the old colors of Trans Canada Air Lines, Air Canada's predecessor (more about this particular aircraft later).

As well as sporting the company logo, airplane tails often get special paint jobs and are fitted with lights to illuminate the company's logo. But the differences between one tail and another go deeper than just the paint job.

Throughout aviation history, designers have built aircraft with some very distinguishable tails. Air Canada's inaugural fleet, some 70 years ago, was no exception. Its Lockheed L10A Electra, the vintage 10-seat "sports car of the skies," had a sleek and shiny brushed-aluminum body and a double-finned tail. The Electra can still be seen today, as part of sightseeing flights in Canada to

Lockheed L10A Electra

raise funds for the Dreams Take Flight program. The Lockheed Super Constellation, which dominated Air Canada's fleet during the '50s, also had a very unusual tail; its triple fins, or vertical stabilizers, made it truly distinctive.

Looks aside, the tail provides directional stability as the aircraft moves through the air and is home to a control surface called the rudder. Functioning much like the rudder on a boat, the rudder on a plane is a vertically hinged wing located on the trailing edge of the vertical stabilizer. It can be steered to the left or right, and pilots use foot pedals on the floor of the flight deck to control it.

Also in the tail section are the horizontal stabilizer and the elevator. As its name suggests, the elevator points the airplane up or down. Pulling back on the control yoke (or the joystick in an Airbus) aims the nose upward. Modern airliners have only one fin at the tail, with the horizontal stabilizer and elevator attached

Lockheed Super Constellation

either high above or level with the main wings. Although the fin's design is now more standardized, its size isn't. On the Boeing 777-300ER, the tail towers over 61 feet (18 m), or 6 stories high. Inside this unpressurized cavern are aluminum ribs, miles of wires, cables, hydraulic lines, and motors that move the large control surfaces. The supporting fuselage also houses the auxiliary power unit, which is an additional jet engine that supplies conditioned air and electricity while on the ground.

With so much going on both inside and out, it's no wonder airplane watchers love to spot different tails. The next time you see the distinctive red maple leaf on the tail of an Air Canada airplane, don't be surprised if you see someone squinting to get a better look at it. Who knows, you may even be tempted to write down the flight identification number yourself.

"It's easy to make a small fortune in aviation. You start with a large fortune."

Airplanes work hard. They must be in the air to make money. For most of the day they fly with minimum ground stops. While parked at the gate, passengers, baggage, and freight are unloaded, and groomers invade the interior, readying it for the next flight. The fueler is hooked up; the caterers open the doors to exchange the old galley for new commissary, and maintenance performs their checks. A new batch of pilots and flight attendants may join this pit stop operation. To see just how busy it is, let's take a look at one day in one airplane's life: FIN 264 on a cold January day.

A Day in the Life

At Gate 20 in Halifax, Flight 601 is being readied by the morning crew for a 6:00 a.m. departure. Bound for Toronto, FIN 264 was carefully looked over by maintenance during the night shift. An outside heater is hooked up, external ground power is connected, lavatories are serviced, the last of the cabin grooming is done, and the commissary is boarded. Some 8000 liters (2110 gal.) of aviation gas are uplifted and 2200 pounds (1000 kg) of lobster destined for overseas sit in the cargo hold. Toronto-based pilots are in the flight-planning office perusing the flight plan; the Halifax-based flight attendants are in their pre-flight briefing.

FIN 264 is an Airbus A319, a typical hard worker in Air Canada's fleet, but what distinguishes it from the others is the original gray paint scheme of Trans Canada Air Lines (TCA).

FIN 264 pushes back on schedule, and its departure time is automatically sent to operations in Toronto when the parking brake is released. The crisp January night caused frost to form on the wings, so it's off to the deicing pad for a quick spray. Ten minutes later, the 30,000 pounds of thrust from each engine pounce the 70-metric-ton (77 ton) airplane into the air.

Daylight gives way to thin, wispy clouds at cruise altitude over northern Maine. Air traffic control, flight dispatch, maintenance control, and Toronto operations are just a few of the many departments monitoring the progress of Flight 601. Once landed at Pearson International Airport, the Toronto-based crew brings FIN 264 to the finish of its first leg of the day. The crew will continue on to Calgary on another aircraft, FIN 224, an Airbus A320.

Unlike the A320, the A319 has no cargo containers so everything, including that precious shipment of lobster, is unloaded manually. FIN 264 has a 50-minute station stop and then it's off to Boston, then back again to Toronto.

Montreal-based pilots and flight attendants will then take FIN 264 from Toronto to Chicago, and back. But before they do, fuel and commissary are restocked again. Also, seat 24C will not recline so the Toronto maintenance team boards the plane during the 40-minute stop.

While taxiing in Chicago, the captain decides to shut down one engine to save on fuel. More fuel is boarded, but this time the quantity is in gallons. The fuel computers quickly and easily convert the gallons to liters, confirming the proper amount. Additional fuel is added because flight dispatch sees a weak snowstorm moving in from Manitoba; Montreal (Pierre Elliott Trudeau) is the backup airport, in case the weather deteriorates in Toronto.

Yet another flight crew boards FIN 264 in Toronto, this time destined for Winnipeg, Manitoba. The aviation forecast has snow starting in Toronto at 5:00 p.m., which shouldn't hamper the 4:00 p.m. departure. On the climb out of Toronto the top of the clouds is some 30,000 feet (9144 m), a phenomenon associated with the incoming weather system. Not a problem for the A319; it cruises at 37,000 feet (11,278 m), in clear and smooth air.

The landing in Winnipeg is a beauty and the passengers tell the pilot so upon leaving, something a pilot loves to hear. During the walk-around by Winnipeg maintenance (see page 39), the team notices one tire is wearing a little. A call is made to Toronto to arrange a replacement for later tonight. At 7:00 p.m., FIN 264 departs on its final leg back to Toronto. Today it has accumulated 12 flight hours and seven landings on its airframe. Since its purchase in 1994, FIN 264 has clocked over 27,000 hours, with close to 13,000 takeoffs and landings.

There is some traffic coming into Pearson so FIN 264 goes into a holding pattern for 20 minutes while Toronto air traffic control clears the buildup. The first officer lands on Runway 23 in strong, gusty winds without a hitch — the A319 can handle 37-mile-per-hour (60 km/h) winds at its side. FIN 264 taxis to Terminal 1. It's nearly 10:00 p.m. as the day unwinds for this workhorse. The pilots do their shutdown checks while maintenance waits by the flight deck door to taxi FIN 264 to a warm, cozy hangar for the night. Early tomorrow morning, it's scheduled to do a Montreal return trip, or "turn" — yet another hard day's work lies ahead.

Just Checking

Safety is a must for any airline, and maintenance is a big part of the safety equation. There are a series of checks every aircraft must go through every day. Everything is recorded. Even aircraft tires have their own log of takeoffs and landings. Sound expensive? It is. But here's one aviation adage that summarizes the situation: "If you think safety is expensive, try having an accident."

Don't be alarmed if you see your pilot walking around the tarmac inspecting the aircraft before departure. He or she is doing a last-minute check for anything out of the ordinary — from fuel leaks and tire wear, to dents and holes caused by the approach of ground vehicles.

The mandatory walk-around is the most basic of an extensive list of maintenance checks carried out by teams of specialists to ensure the airworthiness of an aircraft. Next in ranking is a trip check, which is a walk-around done by maintenance personnel at designated bases. The third type of check is a service check, which must be completed on every aircraft every 48 hours. It includes an inspections review of aircraft systems and lubricant levels. In addition, any problems reported in either the cabin-defect or aircraft-defect logbooks are forwarded to maintenance staff in order for them to repair specific faults prior to each departure.

As an aircraft's flight time and number of landings accumulate, more advanced checks are required, with each aircraft manufacturer having its own requirements. To ensure structural integrity, workability, and reliability, aircraft manufacturers stipulate that these checks must be performed in accordance with government regulations. These scheduled checks also allow any modifications, as

Maintenance on the A340

specified by the manufacturer, to be completed.

Scheduling Air Canada airplanes to receive their appropriate checks can be a major task, especially since the requirements of four aircraft manufacturers must be satisfied. To tackle the job, 3500 Air Canada technical-operations employees are located in two main overhaul bases: Montreal and Winnipeg.

The most extensive of the advanced checks is the H check. To see an airplane being built is quite a sight, but to see one being completely dismantled during an H check is even more amazing. For example, when one of Air Canada's Boeing 767s has accumulated either 72 months in service or 12,000 landings it is rolled into the 1.5-million-square-foot (140,000 m²) Montreal center. There, a team of 250 specialists, divided into three shifts, operate on it around the clock for nearly 47 days.

No system goes untouched. Even the flight deck becomes unrecognizable as the seats are removed, and the panels and floors are lifted up, exposing bundles of wires, cables, instruments, radios, and computers. The cabin, too, is subjected to rigorous inspection. The inner walls and ceilings are dismantled, floors are taken up, seats and galleys are removed. You'd think the airplane was being stripped for parts, but the inspection doesn't end there.

With the aircraft up on jacks, the landing gear is removed, the flight controls (rudder, elevators, ailerons, spoilers, flaps) are taken off, and the engines are separated from their pylons. The pylons (mounts that secure the engine to the wing) are then checked for cracks and other defects. The frame is also closely inspected for possible corrosion, which could be caused by something as innocuous as coffee spilled around the galleys. The wings remain attached, but all of their panels are opened, exposing fuel tanks, wires, and tubing. "Everything is inspected for water and moisture contamination, delamination, cracks, and corrosion," says Ben Minicucci, Air Canada's director of heavy maintenance, from his Montreal office overlooking the huge hangar below.

Each and every component is tested, including slides (passenger chutes used for evacuation), the infamous black boxes, and cockpit voice recorders. In addition, many of the removed components go through rigorous non-destructive testing (NDT), done by X-ray, ultrasound, or conductivity tests. Pressurization checks are also performed and the landing gear is cycled numerous times.

Once all the obligatory checks are done and everything is recorded and documented, the dismantled aircraft is put back together. New galleys are installed, seats are refurbished, a fresh coat of paint is applied, and new carpeting is fitted. This interior makeover gives the impression that the airplane has just rolled off the production line. And, in a sense, it has.

From the basic walk-around to the extensive H check, the intense scrutiny that all aircraft undergo ensures that they run like new.

➤ Number of flight hours before major engine inspections are required for a Boeing 767: 4000.
➤ Number of days required for engine overhaul inspection: 90.
➤ Name of heavy maintenance inspections (the most intense type) for Boeing 767: H check.
➤ When it's time for a B767's H check: after 72 months or 12,000 landings.
➤ Number of days required for a B767's H check: 47.
➤ Examples of H check inspections: flight control overhaul, avionics upgrades, structural inspections (once interiors are completely removed).

Do aircraft need keys to start the engines?
Only small airplanes require a key. An airliner's engines are activated by switches and levers.

Can aircraft auto-land?
Yes, many modern airliners are equipped for auto-lands under specific circumstances.

Can aircraft auto-takeoff?
No, automatic takeoffs do not exist.

Where is aircraft fuel stored?
Many airports have fuel reservoirs embedded in the tarmac.

How do you steer an airplane on the ground?
By a tiller (a hand-operated device that turns the nose wheel).

How do you stop an airplane after landing?
Through the use of brakes, reverse thrust, and ground spoilers.

Now that we've brushed up on some of the intricacies of an airplane
— like building and painting an aircraft — and examined some of
its components and its maintenance, let's go flying!

CHAPTER THREE

Before-Takeoff Checklist

From the Gate into the Air

All aboard! Toronto to Hong Kong!

As promised, I'll be taking you on a virtual flight from Toronto non-stop to Hong Kong, revealing tons of interesting facts and considerations along the way. This is Air Canada's longest flight. Coming in second, by only five minutes, was New Delhi, India, direct to Toronto, which traversed 13 countries.

Check-in time for passengers is two hours before departure, but a multitude of behind-the-scenes teams are already working on the flight by that time. About three hours before departure, flight dispatch completes the flight plan, which is generated by software from Germany's Lufthansa, one of the largest airlines in the world. Included in this lengthy flight plan are the routing, altitudes to be flown, fuel burns, pertinent notes on airports along the route, weather, aircraft snags, and so on. Even though we don't have direct contact with flight dispatch, they are only a phone call or data link away. They watch every move we make; part of their duty is following flights. Known as "captains on the ground," they are the second-highest paid group (next to pilots) at the airline.

Captains on the Ground

The lights are always on in an office tower overlooking Toronto's

Pearson International Airport. Here an elite group of dispatchers works around the clock, planning and monitoring more than 650 flights a day. This is the nerve center of Air Canada's operations. Phones are constantly ringing, and the energy level is high as dispatchers analyze weather patterns and make crucial decisions about when and where airplanes will fly.

Brian MacCourt, a 35-year veteran with Air Canada now retired, was the chief of flight dispatch and took me on a tour of the operation. He moved easily amid the networks of computer terminals and communications equipment. After all, he helped design this high-tech workplace. The control center has 16 dispatch stations. Each one is identical, with four computer terminals and a communications system capable of storing numerous radio frequencies and 5000 telephone numbers. The on-duty chief dispatcher sits on an elevated platform as display terminals provide the whereabouts and up-to-the-minute status of every airplane in the fleet.

Each station regularly has nearly a dozen (sometimes over 20) programs up and running at a time. One of those programs may be the feed from air traffic control making the details of every flight only a mouse click away. The dispatcher can query everything from a flight's altitude and ground speed to its routing and arrival time. Once the aircraft pushes back from the gate, the dispatcher, often referred to as the "third pilot," follows the flight to its destination. That means there are no regularly scheduled breaks for lunch or for coffee — the dispatcher's breaks depend on the flight plans. Whenever a dispatcher has to leave his or her station another dispatcher is assigned to fill in.

When dispatcher Esther Eivinsen, on the Maritimes/Ottawa/Montreal desk, receives information about a possible bird-strike incident in New Brunswick, she fires off a query to the pilot. After

a few tense moments, she receives a data link message that the aircraft is fine. She then quickly briefs the duty chief. At this point in

Dispatcher Esther Eivinsen working the Maritimes/Ottawa/Montreal desk

Toronto's flight dispatch, "where the lights are always on"

her shift, the workload is fairly light; there are only five aircraft on her screen. A dispatcher can handle up to 40 flights per shift.

The second terminal at each station may display the latest satellite pictures, weather maps, and radar images. Computer software allows dispatchers to superimpose the weather information onto the path of an airplane. If a potential problem lies ahead, they can send a message directly to the pilot by data link.

The dispatcher may have the flight planning program armed and ready on the third terminal. Software called Lufthansa Integrated Dispatch Operation, or LIDO, is used to calculate the best route for each flight. Among the factors considered are the forecasted winds, available altitudes, weather conditions, ride reports from previous flights, and aircraft type. With flight plans containing so much information, a printout for an overseas flight plan can easily fill 30 pages. By optimizing flight plans, a dispatcher can also conserve a great deal of fuel. As mentioned, LIDO calculates fuel consumption to within 100 kilograms (220 lb.) and arrival times to the exact minute. This is an amazing feat, considering an Airbus A340 may carry more than 90,000 kilograms (198,416 lb.) of fuel for a flight lasting 13 hours.

The fourth terminal will have a gamut of other programs running, providing information from curfew approvals, runway conditions, and phone numbers to the latest sports scores for pilots to pass along to their passengers.

Surveying the room full of dispatchers at their terminals, Brian MacCourt says each of the 67 men and women on his team has been carefully trained to carry out the complex tasks of dispatching aircraft. To obtain their certification, dispatchers study a wide range of subjects, including meteorology, air law, communications, and air traffic control. Even after they are certified, dispatchers undergo annual tests and take extra classes to stay current. "Our team of flight dispatchers is the best in the world," says MacCourt. "That's why we're in demand to provide other airlines with our knowledge and expertise."

✈ ✈ ✈

Flying from Toronto to Hong Kong will require four pilots as it is deemed an ultra-long-haul flight. We may well meet for the first time at flight planning, so it's not uncommon to begin with introductions. We retrieve the flight plan, print the latest weather charts, correspond with dispatch for ride reports, all in the company's busiest flight planning center. Because our routing will take us over the North Pole, "space weather" conditions must also be considered.

Over-the-Top Weather

Until recently, the only regularly scheduled flights around the North Pole were handled by a roly-poly man with a white beard and red suit; even then there was only one flight a year. But the opening up of former-Soviet airspace at the turn of the 21st century created new opportunities, and today many airlines are launching flights over the top daily. But this newfound flight path comes with many restrictions and new meteorological considerations.

Why the sudden urge to take it over the top? Warren Lampitt, general manager, technical programs, flight operations at Air Canada explained, "It knocks off at least one hour of flying time." And on a four-engine jumbo jet one hour equates to thousands of dollars in fuel savings. (The A340-500's thirst for fuel is about 8000 kg/h, or 17,640 lb/h.) "Because less fuel is required, more baggage and cargo can be added, and no fuel stop is needed — all adding to the bottom line," Lampitt continued. Additionally, turbulence is less prevalent because jet streams are corkscrewing the globe further south and there are no weather fronts to contend with. And, obviously, crew-duty time is also lessened. It may not seem like a big issue, but it easily enters the equation if a less-productive routing is flown.

Space Weather Hurdles
North Pole flights do present some unusual obstacles, especially as

it concerns space weather. Space weather, as used here, is defined as the conditions created on the Earth by activity on the surface of the sun. The potential impact of electromagnetic and solar radiation has been categorized by the National Oceanic Atmospheric Administration, or NOAA (see sidebar, page 48). Cosmic radiation levels are an important space weather concern. NOAA ranks these levels on a severity scale from one to five. S1 and S2 allow for a safe journey, whereas an S5 is equal to about 100 chest X-rays. Forecasts of levels of S4 and S5 prohibit polar flights, with S3 imposing lower altitudes or a more southerly polar route.

Aircrew and passengers run a slightly higher risk of cosmic radiation at higher flight levels, and this risk increases toward the poles. (Four factors affect the potential dose of cosmic radiation: latitude, altitude, flight duration, and solar activity.) The atmosphere offers less-inherent protection at higher altitudes, and protection also lessens toward the poles. The atmosphere is thinner at the poles, and the Earth's protective magnetic field also decreases further still near the poles. Incoming cosmic radiation particles are deflected by the Earth's magnetic field, and in general, radiation shielding is greatest at the equator and decreases as one goes north or south.

Another element of flight impacted by space weather is radio reception. Again, NOAA broadcasts a five-level range of severity. Some high-frequency (HF) aircraft radios work by bouncing transmissions off the ionosphere (the layer of the Earth's atmosphere ionized by solar radiation), which allows for long-distance communications. An R5 rating means radio communication would not be possible for hours, and the aircraft would not be able to communicate with air traffic control. Luckily, Future Air Navigation Systems (FANS) work through the use of satellites, lessening the reliance on HF; FANS played a major role in making polar flights a reality. Warren Lampitt notes that, due to difficulties with SATCOM (satellite communication) data link and HF voice communications at high latitudes, Air Canada implemented HF data link, which bounces radio waves for hundreds to thousands of miles along the ionosphere, on the B777 fleet when it entered service this year.

Yet another feature of space weather is the geomagnetic storm scale, which measures worldwide disturbances of the Earth's magnetic field. It, too, is ranked one to five. A G1 rating indicates slight power-grid fluctuations and minor impact to satellites, whereas a G5 is extreme and denotes possible power-grid collapses, damaged transformers, and radio blackout in many areas for one to two days. During a G5, satellite navigation and communication unreliability coupled with possible ground power outages would cancel polar-route flights.

NOAA's Space Environment Center (SEC) operates a worldwide network of sensors that continuously observe conditions between the Earth and the sun. Its website (www.sec.noaa.gov/SWN/index.html) offers excellent real-time information on electromagnetic and solar radiation.

Solar Radiation Storm Scale
S5 (Extreme): High radiation hazard to commercial jets (equal to 100 chest X-rays), loss of some satellites, no HF communications in polar regions.
S4 (Severe): Radiation hazard to commercial jets (equal to 10 chest X-rays), satellite star tracker orientation problems, blackout of HF radio at polar cap for several days.
S3 (Strong): Radiation hazard to jet passengers (equal to 1 chest X-ray), temporary upset to exposed satellite components, degraded HF at polar cap.
S2 (Moderate): Infrequent satellite event upsets, slight effect to polar cap HF.
S1 (Minor): Small effect on HF radio in polar region.

Radio Blackout Scale
R5 (Extreme): Complete HF radio blackout on the entire sunlit side of the Earth for a number of hours, navigational outages on sunlit side for many hours.
R4 (Severe): One- to two-hour HF blackout on sunlit side of Earth, minor satellite navigation disruptions.
R3 (Strong): Wide area of HF blackout, loss of radio contact for mariners and en-route aviators for about an hour, low-frequency navigation (LORAN) degraded.
R2 (Moderate): Limited loss of HF radio, some low frequency navigation signals degraded.
R1 (Minor): Minor degradation of HF, minor low-frequency navigation signal degrade.

Geomagnetic Storm Scale
G5 (Extreme): Power grids can collapse, transformers are damaged, powerful electric charge can threaten satellites and spacecraft operations, HF radio blackout in many areas for one to two days, low-frequency radio out for many hours, aurora seen as low as the equator.

G4 (Severe): Voltage stability problems in power systems, satellite orientation problems, induced pipeline currents, HF radio propagation sporadic, low-frequency radio disrupted, aurora seen as low as the tropics.

G3 (Strong): Voltage corrections required on power systems, false alarms triggered on protection devices, increased drag on satellites, low-frequency radio navigation problems, aurora seen as low as mid-latitudes.

G2 (Moderate): High-latitude power systems affected, drag on satellites affect orbit, HF radio propagation fades at higher altitudes, aurora seen at latitudes of 50 degrees.

G1 (Minor): Slight power grid fluctuations, minor impact to satellites, aurora seen at high latitudes (60 degrees).

Adapting to Space Weather

What can be done about space weather? Nothing. Airlines using polar routes have adopted the policy that flights will not be conducted if solar radiation, radio blackout, or geomagnetic storm activity is at level four or five. Solar radiation at level three requires polar flights to be conducted at FL 310 or below. ("FL" stands for flight level, or height above sea level. FL 310 means 31,000 feet above sea level.)

As well, there are four polar routes to choose from, with Polar Route 2 being the closest to the pole, about 60 miles (97 km) away. (No route travels directly over the pole.) Hours before each polar flight, flight dispatch determines whether space weather is deemed safe. Sometimes a safe flight can only be assured by varying the route or changing the flight level.

Frigid Factors

Another weather contender during our flight to Hong Kong is the extreme cold found in northern Canada and Siberia, which could potentially freeze the fuel. Flights into areas of –65°C (–85°F) must be restricted to 90 minutes. Depending on the aircraft, engine type, and jet fuel type, the fuel boarded may be analyzed and the actual fuel-freeze point determined. Flight dispatch may data link this actual fuel-freeze temperature to the flight after it is airborne.

On a recent flight of mine, this very problem came up as the jet fuel cooled to below –40°C (–40°F), triggering a message in the fuel-temperature-monitoring system. The fuel boarded had a threshold of –47°C (–53°F), making immediate action unwarranted.

If the conditions had persisted, procedures state we would have had to either descend into warmer air or speed up. Descending burns more fuel, as does increasing speed, but it is less effective. Luckily temperatures were forecasted to warm up and they did.

Yet another consideration is the availability of suitable airports in case of a serious medical situation or other emergency, particularly in Arctic winters. Two Arctic survival suits, along with other environmentally appropriate clothing, are on board in case we have to exit the airplane to coordinate services after landing. (Rest assured the junior pilot will be delegated this task.) One airport in close proximity of the transpolar route is Tiksi, in Russia. Briefing notes highlight the fact that it lies in the coldest region of the northern hemisphere, with temperatures reaching as low as −71°C (−96°F). A recent Air Canada memo suggested only landing at such airports in perilous situations.

Passengers are taken aback when they realize our routing will take us north because, after all, Hong Kong is west of Toronto. When you join two dots on a sphere, or two airports on Earth, the shortest distance is called a great circle. Try it yourself. If you take an atlas and join Toronto to Hong Kong with a string, the route direction is southwesterly over California. Now connect Toronto to Hong Kong with the same string using a globe. You'll notice it makes sense to head north instead of southwest for this trip. On this flight you would glimpse Hudson Bay and the frozen tundra of Siberia instead of California's sandy beaches. However, because of reasons we'll explore below, flights are rarely as the crow flies.

Plotting a Path

Aviation maps in many in-flight magazines depict airline routes as clean, curved lines that gracefully arch from one point to another. They provide a clear connect-the-dots picture of which cities the airlines serve, but the true picture is very different. If one were actually to combine all the airplanes' flight plans, the lines wouldn't be smooth and gently curved; they would look like a vastly intricate spider's web. Routes twist and diverge in all directions, zigzag from place to place, and often intersect with other routes. Many variables come into play when plotting out routes, which at first glance often appear indirect and puzzling. If only the airplanes could travel in the smooth lines on the maps — it would all be so much simpler.

While the shortest distance, and often the fastest way, between two points on the ground is a straight line, this isn't true when traveling along windy air routes. A favorable wind, or one working against the airplane, can make a big difference to the ground speed (the speed at which the aircraft travels over the ground), so dispatchers must carefully consider the direction of the winds when selecting a route. As a result, sometimes the best route to where you are going zigs (or zags) to line up the airplane's flight path with the direction of the prevailing winds, or to avoid flying right into those winds. Wind speed generally increases with height, with values sometimes exceeding 250 miles per hour (400 km/h), and the wind's direction is constantly shifting. As a result, the routing over the North Atlantic, for example, changes daily. The routes handling the daytime westbound traffic from Europe tend to be more to the north, to avoid the headwinds, while the eastbound night traffic tends more to the south.

Weather also dictates the direction of airways. Thunderstorms can act as roadblocks, putting a detour in the flight's route. For example, a Vancouver-bound flight from Toronto may head into American airspace to avoid thunderstorms over the Prairies.

The capabilities of the airplane itself also dictate what routes it is permitted to travel. At one time, a two-engine jet had to remain within an hour from a suitable airport while traveling over the Atlantic or Pacific oceans. For the Atlantic route this meant hugging close to Greenland and Iceland. This posed major restrictions on ocean crossings. Today, Air Canada's large two-engine jets — the Boeing 767, the B777, and the Airbus A330 — are permitted to operate three hours from land, thanks to the increased reliability of modern-day jet engines.

Some restrictions to international airways are political. Many countries — such as Iraq, the Russian Federation, and China — strictly limit or forbid airliners to fly in their airspace. It was not too long ago that talks between NAV CANADA and the Federal Aviation Authority of Russia led to relaxed airspace restrictions, allowing Air Canada to route flights over the North Pole and other parts of the Russian Federation. This saved time and improved service to Asian destinations, permitting, for example, our non-stop flight from Toronto to Hong Kong, which had nearly one hour cut from the flight time.

Some routes handle air traffic for specific directions, whereas most routes handle traffic both ways. Unlike having two separate lanes, aircraft keep a safe distance by varying altitudes according to direction. Many of these route intersections are named after local places; others have more random or peculiar names. My favorite is the waypoint called Crazy Woman in Wyoming. Because of sophisticated navigation systems, many waypoints are now just that, points in space with specific latitudes and longitudes. These markers allow air traffic controllers to provide better traffic flow, and many waypoints near airports have assigned altitude restrictions. The majority of the names are now computer generated. (These peculiar words would come in handy for a game of Scrabble. For example, one approach into Toronto Pearson has the pilot navigating to the following points: XEXAX, ERBUS, ITROX, and DULPA.)

With sophisticated inertial navigation systems and the onset of satellites, routes are becoming more direct. Frequently pilots will request a more direct routing from air traffic control to save time. As long as there is no conflict — such as other air traffic or active-practice military zones that won't allow overhead commercial flights — the controllers will try to accommodate the request. So, with once-forbidden airspace opening up, engine technology advancing, and the highly accurate navigation systems in use, restrictions on where an airplane can fly are becoming less stringent. Who knows? Maybe one day flying in straight lines from one city to the next will be the best way to go.

Cracking the Code

When booking a flight, reading your trip's itinerary, or looking at the tags on your checked luggage, you'll notice three-letter codes that identify airports. Sometimes it's clear which airport the three letters identify, but more often than not they are an illogical abbreviation. For many passengers it's one of those mysteries of travel. Below I'll try to dispel some of the secrecy and unravel the codes. Not to worry, it's not as complicated as *The Da Vinci Code*.

You're off to Chicago and you notice three mystifying letters —

ORD — on your checked baggage. Why the confusing city code? Why not, say, CHI? History has a lot to do with it. Airport codes are designated by geographical locations, names of airports, and personal tributes. As such, DEN is for Denver, Colorado, and JFK honors President John F. Kennedy (for the airport in New York that bears his name). ATL, the code for the world's busiest airport, Atlanta, is obvious. So why are some codes, such as CDG for Charles de Gaulle, easy to crack, while others, such as YXE for Saskatoon, are real brainteasers?

Years ago, the National Weather Service devised a two-letter identification system to keep a handle on weather throughout the United States. When aviation was in its infancy, airlines simply adopted that system. However, major expansion meant towns without weather stations needed codes as well, and so IATA, the International Air Transport Association, created three-letter identifiers for airports around the world. Canadian weather offices associated with an airport used the letter Y, which makes it easy to identify a Canadian airport, but difficult to remember the individual codes. Some are easy to figure out: Vancouver is YVR, YWG identifies Winnipeg, and YQB designates Quebec City. But Toronto's Lester B. Pearson Airport, the country's busiest airport, is the not-so-obvious YYZ.

Incidentally, Chicago's ORD is named for Orchard Field, and that former airstrip's moniker is a tribute to pilot Lt. Cmdr. Edward O'Hare who, in World War II, was the U.S. Navy's first flying ace and a recipient of the Medal of Honor.

To make things more confusing, the International Civil Aviation Organization (ICAO) implemented a four-letter identifier for each airport. These codes are used for flight planning, aircraft navigation computers, and weather info. So you and your luggage may be off to LHR, but I input EGLL into the flight computer. Want to unravel this code? It's E for northern Europe, G for Great Britain, L for London-control airspace, and L for London, Heathrow airport. Canada and the United States use C and K respectively for ICAO codes, so YOW (Ottawa) becomes CYOW and BOS (Boston) is KBOS.

- -

Airport codes are need-to-know information, and there are many websites that can help you bust the codes (www.world-airport-codes.com is a good one). Here are the codes for the 10 busiest airports (by passenger traffic):

1. Hartsfield-Jackson Atlanta International Airport
Atlanta, Georgia, United States ATL/KATL

2. O'Hare International Airport
Chicago, Illinois, United States ORD/KORD

3. London Heathrow Airport
Hayes, Greater London, United Kingdom LHR/EGLL

4. Tokyo International Airport (Haneda)
Ota, Tokyo, Japan HND/RJTT

5. Los Angeles International Airport
Los Angeles, California, United States LAX/KLAX

6. Dallas–Fort Worth International Airport
Dallas–Fort Worth, Texas, United States DFW/KDFW

7. Charles De Gaulle International Airport
Roissy, Tremblay-en-France, France CDG/LFPG

8. Frankfurt International Airport
Frankfurt, Hesse, Germany FRA/EDDF

9. Amsterdam Schiphol Airport
Haarlemmermeer, North Holland, The Netherlands AMS/EHAM

10. McCarran International Airport
Las Vegas, Nevada, United States LAS/KLAS

Here's why you should always be nice to your check-in agent . . .

   An irate passenger checking in for a flight confronted a ticket agent and demanded prompt service. His rudeness and gruff conduct never seemed to faze the agent, as she remained coolheaded and calm the entire time. Finally, after it was all over, another agent, who had noticed this tense situation, asked the cool ticket agent how she remained so calm. The agent just smiled and said, "The passenger may be going to New York but his bags are off to Detroit."

During my first full-time flying job,
I loaded the cargo myself

As you settle into your seat you'll notice machinery and ground crew loading the aircraft with baggage and cargo. You may even see an armored vehicle loading money (no, it's not payday for the pilots) and valuables into the belly. Most aircraft's cargo areas are accessed from the right, with much of the load boarded in containers. The smaller commuter jets' and turboprop aircraft's cargo holds are accessed toward the back, on the left.

Baggage handlers load more than luggage into the belly of an aircraft. Without knowing it, passengers may be sitting over racing cars, polar bears, or tons of green onions! Passenger flights carry cargo that range from shipments of Israeli flowers and vintage French wine, to live lobsters, racing horses, sail boats, and zoo animals like seals and lions. It's all part of a long cargo tradition. For Air Canada, that tradition dates back to their very first flight, a short jaunt between Vancouver and Seattle in 1937 that carried two passengers and a consignment of mail.

In those days, post office regulations required that all mail be carried in a closed compartment to prevent tampering. To comply, shipments were locked in the washroom at the rear of the cabin; a solution passengers must have found a little frustrating. Years later mail compartments were installed below the flight deck. Air Canada now carries nearly 77-million pounds (35-million kg) of mail annually; Canada Post is the airline's largest cargo customer.

In the past, Air Canada operated aircraft that transported cargo exclusively. Now passenger airplanes carry all the airline's cargo shipments. Though no longer part of Air Canada's fleet, the Boeing 747-400 Combi — a specially built airplane that is half passenger cabin and half cargo bay — can carry up to 85,323 pounds (38,702 kg) in its back section, more than twice the weight of its full load of 265 passengers. More commonly, cargo is carried in the belly of pas-

Loading precious cargo onto an Air Canada passenger flight

senger aircraft: the airplane you're flying in could contain 44,000 pounds (20,000 kg) of asparagus or even a rare Asian elephant!

On November 7, 1997, Air Canada did, in fact, ship an elephant, a 3993-pound (1810 kg) behemoth named Emmett. Born at the Burnet Park Zoo in Syracuse, New York, six-year old Emmett was needed in London, England, where three childless female elephants were awaiting his arrival. A specially built crate and extra-large forklift were required to help speed Emmett toward his paternal responsibilities. Emmett traveled with two zookeepers and a veterinarian in case of emergency, but everything went smoothly — thanks to a supply of his favorite treat, jelly beans.

Pilots are always notified about live animals on board. Most pilots never venture into their aircraft's cargo holds though a few have been known to take a peek at a lion. Cargo holds are always pressurized but are unheated in aircraft like the Airbus A319, A320, and the Canadair Regional Jet. As a result, live animals cannot be shipped on most of these aircraft during the winter.

Why the restrictions? Air in the cargo hold is recirculated throughout the airplane, so the heated air from the cabin keeps the hold above freezing. However, at high altitudes the outside temperature is always very cold, while at mid-level altitudes temperatures fluctuate 10 to 15 degrees Celsius (depending on the season). In extreme weather, Air Canada always evaluates the shipping of animals on a case-by-case basis.

Next time you watch baggage being loaded, look more closely. You might see an armored vehicle deposit some priceless article for safekeeping, or witness the loading of rare museum artifacts on their way to a faraway exhibition. Or maybe your overzealous aunt has shipped your Christmas gifts a few weeks early. You might even hear a rumble that doesn't sound like an engine — it's an elephant trumpeting for more jelly beans in the hold.

Fully Loaded

➤ Cargo capacity of Boeing 747 Combi: 85,323 lb. (38,702 kg).

➤ Cargo capacity of Dash 8-100 series: 665 lb. (302 kg).

➤ Location of flight attendants' resting area for long-haul flight: cargo hold.

➤ Name of resting area: LDMCR (lower deck mobile crew rest).

➤ Name of Air Canada's new state-of-the-art temperature-controlled cargo container: AC Cool.

➤ Contents suitable for AC Cool container: perishables, pharmaceuticals, fine art.

➤ Temperature range of AC Cool cargo container: –20°C to 20°C (–4°F to 68°F).

➤ Air Canada's biggest and oldest cargo customer: Canada Post.

➤ Amount of mail transported annually by Air Canada Cargo: 77-million lb. (35-million kg).

➤ Weight of cherries transported during peak cherry season (June–July): 5.5-million lb. (2.5 million kg).

➤ Size of new Air Canada Cargo facility in Toronto: 287,000 sq. ft. (26,660 m^2).

➤ Name of the area in this new cargo facility for large animals: animal hotel.

➤ Interesting species transported in cargo: gorillas, lions, koalas, elephants, Komodo dragons.

➤ Most frightening cargo witnessed by lead station attendant Perry Mulhall: two hissing black panthers.

➤ Official carrier for Toronto Zoo: Air Canada.

Aviation has its fair share of rules, and sometimes I wonder if all of our many cabin announcements are necessary. After all, if a person doesn't know how to fasten his or her seat belt, it will probably take more than an announcement to set things straight. After spending many hours in the back of an airplane and logging as many hours as a frequent flyer, I make sure I'm armed with a crossword or Sudoku to help me get through the PA chatter. But all these announcements have a purpose: maintaining passenger and aircraft security.

To quote from one flight attendant's announcement, "There may be 50 ways to leave your lover, but there are only 4 ways out of this airplane."

Why All the Rules?

From the moment you step into an airplane, rules and regulations dictate practically everything you do. You have to remain strapped into your seat, turn off your iPod, and even take away some of your children's toys. Many of these rules may seem meddlesome and even silly at times, but they help ensure the safety and comfort of passengers, and they're based on the most recent research into aviation safety.

Fairly recently, most airlines began asking customers to keep their seat belts fastened while seated throughout the flight. The rule was implemented following an incident on a Boeing 747 that encountered severe turbulence east of Tokyo in 1997. One passenger died and many were flung around while seated with their seat belts unfastened. Even though pilots can often anticipate areas of turbulence, situations still arise where roughness escalates rapidly. In fact, flight turbulence is the leading cause of injuries to passengers and flight attendants. Keeping seat belts fastened is one way to help minimize the risk. All air carriers are obligated to make mandatory seat-belt announcements before the flight gets airborne, as well as an announcement that carry-on luggage, which includes purses, brief cases, jackets, and laptops, must be safely stowed. It's the law!

Another regulation is the prohibition against using iPods and other personal listening devices during taxi, takeoff, and landing. The last thing any airline wants to do is deny customers the pleasure of listening to their favorite music, but the rule was implemented in 1997 after it was found that these devices prevent passengers from clearly hearing important announcements and flight-attendant commands. Misinterpretations of these instructions caused confusion during debarkation. An airplane's own headsets are not a problem since the PA system overrides them automatically.

You may view aviation rules as a little strict, but just think how much safer roads would be if we drove our cars with as many rules and regulations.

Air Canada has a fairly extensive list of electronic items that must be switched off and safely stowed into bags throughout the flight. These items, which include radio-controlled toys, printers, laser pointers, and video cameras, emit frequencies that may interfere with the navigation and other systems of some aircraft. Don't worry — implanted medical devices, hearing aids, and watches are safe.

Members of the flight crew also frequently have to remind passengers to switch off their cell phones during the flight. As with other electronic devices, there have been documented cases where cell phones interfered with an airplane's navigation instruments.

What's more, cellular phone use while airborne can cause substantial jamming of cellular repeater stations en route. By transmitting to many ground stations at once, cell phones can cause significant interference to the ground system.

Air Canada has recently relaxed the cell phone prohibition slightly, permitting their use during extensive ground delays while the aircraft is parked at the gate with the boarding door open, or while away from the gate with the engines shut down. The captain will make an announcement to inform passengers when these special conditions apply.

One of the more controversial airline rules is the 13-year-old prohibition against smoking on board flights. Many people, including smokers, applaud this rule because it improves the overall air quality on board. However, there are those who find that the smoking ban creates a real hardship — especially on overseas flights. Every week, people are caught lighting up in the washroom. The built-in smoke detector tips off the crew by sending an automatic signal to the flight deck. Rule breakers will receive a fine.

In many parts of the world, for many years, rules such as a smoking ban were unheard of but most airlines have gradually "butted out." In fact, some airlines will be replacing the standard "no smoking" sign with a "no mobile telephone" sign. Instead of seeing a cigarette crossed out, you'll see a crossed-out mobile phone that will be illuminated from takeoff until the plane has reached a high-enough altitude to ensure there is no interference with mobile networks on the ground.

In aviation lingo, the flight attendants are known as the "back end," and the pilots the "front end."

A 16-hour flight can be quite draining for travelers. Imagine what it must be like for the flight attendants working these long-haul flights.

An hour before scheduled takeoff, the flight attendants are already busy preparing for the trip. They review safety and emergency procedures, check that all emergency equipment is available and functional, and verify that the necessary supplies have been loaded onto the airplane.

Each of the attendants is assigned to a specific location; for a long-haul flight aboard an A340-500, this means 14 attendants among the four galleys or the cabin. (Air Canada has over 7000 flight attendants, so introductions are frequently necessary.) Once the safety and emergency procedures review has been completed, everything in the back has been checked, and the names and positions of the cabin crew are noted, the head flight attendant, now known as the service director, takes the cabin safety equipment checklist to the flight deck.

The members of the cabin crew then greet and assist passengers as they board and ensure that seat belts are fastened and luggage is safely stowed, all while carefully supervising any unaccompanied minors. The service director also delivers the final passenger count to the flight deck and lets the pilots know when the cabin is secure for takeoff.

During the pushback from the gate and taxi to the runway, flight attendants used to stand in the aisles and demonstrate safety procedures. However, videotaped information is now used on many aircraft, including the A340-500, which cuts down considerably on the pre-takeoff workload, but you still see flight attendants giving live briefings on smaller aircraft or when the system is not working. At the one-minute-before-takeoff signal from the flight deck, passengers and cabin crew must be seated and strapped in. From that time on, until the seat belt sign is turned off, flight attendants focus on flight progress and are ready for any event such as an aborted takeoff.

When the airplane climbs through 10,000 feet (3048 m) and the seat belt sign is turned off, the flight attendants swing into action. They handle two full meal services and distribute countless beverages and snacks during a long-haul flight. They also hand out headsets, pillows, and blankets and, if required, look after unaccompanied children and passengers with special needs.

The most common questions flight attendants used to hear were "Where are we?" and "How much longer?" Today, a number of aircraft are equipped with a map display that provides those answers, leaving more time to attend to passengers' needs.

Times have certainly changed since 1938, when Lucile Garner and Pat Eccleston began flying with Trans Canada Air Lines. The early stewardesses had to be registered nurses, single, and between 21 and 26 years of age. During World War II, when transcontinental flights were introduced, men joined their ranks as pursers. In the 1970s, men and women were given the same designation: flight attendant. Today they work together to ensure our passengers receive the highest levels of safety and service in the air.

As the flight descends to 10,000 feet (3048 m), about 10 minutes before landing, the seat belt sign comes on again, signaling it's time to pick up any remaining trash and collect the headsets. Another trip ends smoothly thanks to the tireless work of the flight attendants. When we taxi to the gate and set the parking brake, they take their positions to bid passengers goodbye. Tonight they will have a well-earned rest at a hotel. Tomorrow these unsung heroes will do it all over again on their flight back home.

Overheard from a customs officer: "If I ever remarry, it will be to a flight attendant, because it's amazing what one can buy for just 20 dollars."

One flight attendant's embarrassing moment: She noticed a business-class male passenger making rhythmic movements under his blanket. She couldn't believe his blatant maneuvers in public. After seeing enough she confronted the man only to find out he was cleaning his glasses.

Flight attendants' pilot jokes:
What does a pilot use for birth control during a layover?
His personality or his layover clothes.

How can you tell the airline pilot at a cocktail party?
You don't have to, they'll tell you.

Pilots' retort:
What's the difference between a jet engine and a flight attendant?
A jet engine stops whining at the gate.

Ready for takeoff

The flight is now closed up. We complete our before-start checklist and commence pushback. The "out time" is transmitted to operations via data link as soon as the captain releases the parking brake. The aircraft is pushed with a tug or tractor connected via a tow bar, and more modern equipment lifts the nose wheel up into a cradle. The "lead" ramp handler is plugged into a receptacle on the lower portion of the nose allowing two-way communication with the flight crew. Once the aircraft is in a safe area, so as not to cause harm from its jet blast, clearance is given for engine start. With a combination of buttons and levers, the captain starts the four engines during the push. No keys are required. The tow bar is disconnected, a "thanks for the push" is given to the lead, and a taxi clearance is requested from air traffic control.

View From the Ground

Ever feel as though you are being driven rather than flown to your destination? Taxiing to and from the runway can be a lengthy process, as some large airports can be congested. A good way to pass the time is to look out the window and identify the many signs alongside the taxiways and runways.

The large letters you see en route to the runway denote a specific taxiway. Larger airports have numerous taxiways, so pilots have airport diagrams in their route manuals to aid them. You can also watch out for letters, designating the aircraft registration, that appear on the tails of airplanes. You may recall that Canadian aircraft registrations start with C, and American ones with N.

Numbers are also used to represent different runways. A number 24 for example, represents a runway that is oriented 240 degrees to magnetic north. Busier airports have parallel runways, so L or R is added to denote left and right. Numbers are also painted on the ground at the start of runways to avoid confusion.

For a pilot, getting around on the ground can sometimes be more of a challenge than navigating in the air. To ensure that only one aircraft occupies an area at any given time, a taxi clearance is required from both the ground controller and the controller handling the runway's landings and takeoffs. Large airports also have apron or ramp controllers who handle all aircraft leaving and entering the gates.

Taxi clearances are strictly observed. When a controller tells a pilot to "hold short," the pilot must stop and wait at a designated point. He or she also repeats "hold short" to eliminate any confusion. The pilot must also come to a halt where the taxiway meets the runway; this is marked by solid and broken white lines called hold lines. Hold signs placed near the yellow lines serve as a backup, especially for when visibility is poor. Pilots steer the airplane with a tiller, a small hand lever that functions only on the ground. Pedals operate the rudder and nose wheel. The control column or joystick (depending on the type of aircraft) serves no purpose on the ground.

A solid yellow line is painted down the middle of each taxiway. Unlike on the highway, where cars stay to one side of the painted line, on taxiways pilots make sure the airplane's nose wheel is directly on the centerline at all times in order to eliminate any chance of collision. This is a challenge when learning to taxi the A340. The flight deck pivots some 14 feet (4.3 m) ahead of the nose wheel, so the pilot must oversteer the yellow lines when turning. Learning how to turn this airplane on the runway in a simulator is also a challenge. It can't be done on a 150-foot (46 m) wide runway, and when executing the turn on a wider 200-foot (61 m) runway, pilots must place the flight deck (i.e., themselves) well over the grass, something that takes a little getting used to.

Taxiways are lit with blue lights set 200 feet (61 m) apart, while runway lights are white. More advanced airports have green lights embedded along the yellow line to aid the pilot at times of low visibility. The bumps you feel when your plane races down the runway are white lights embedded in the center of the runway.

The tops of nearby buildings and hills are lit with red lights, while obstructions 500 feet (152 m) above ground or higher (the CN Tower, for example) should be equipped with strobe lights. At the request of the pilot, the intensity of runway lights and approach lights can be adjusted all the way up to strength five. Incidentally, in case of sudden power outages, airports have backup generators that become operational within 15 seconds.

Runways at airports that handle jet aircraft are generally 150 to 200 feet (45 to 60 m) wide, while their lengths vary from 5000 to 12,000 feet (1500 to 3700 m). Taxiways are narrower than runways, which is why the wingtips of some aircraft, such as the A340 and the B747, may hang well over the grass. Runways are constructed of concrete and asphalt over a thick gravel bed. With 400,000 kilograms (881,850 lb.) of aircraft touching down at an approach speed of 140 miles per hour (225 km/h), the runway has to be strong and secure.

Lighted windsocks at each end of the runway indicate the direction of the wind; these assist pilots in fine-tuning their landings and takeoffs. A windsock blown into a horizontal position indicates a moderate wind of 15 knots (17 mph or 28 km/h). Other sights you'll notice near runways include equipment that determines visibility, deicing bays and, of course, the control tower.

Now that you know what to look for the next time you're taxiing to or from the runway, you'll have something to keep you occupied. With any luck, the mandatory one-minute-before-takeoff announcement will come sooner than you think.

- -

The Rundown

➤ Typical width of runways: 150–200 ft. (45–60 m).
➤ Maximum slope of a runway for an airliner: +/−2°.
➤ Main surface of a runway: concrete, asphalt, or both.
➤ Distance that separates runway lights: 200 ft. (61 m).
➤ Color of runway lights: white.
➤ Color of taxiway lights: blue.
➤ Direction runways are oriented to: magnetic north.
➤ Exception to this rule: in the far north, runways are oriented to true north.
➤ Meaning of the numbers at the beginning of a runway heading: Runway 24 means it is oriented to 240° magnetic north.

- Reason runway numbers change over the years: magnetic north moves.
- Location of the longest commercial runway in North America: Denver International Airport (DEN).
- Length of Runway 16R/34L at DEN: 16,000 ft. (4877 m).
- Location of the longest commercial runway in Canada: Calgary International Airport (YYC).
- Length of Runway 16/34: 12,675 ft. (3863 m).
- Interesting fact for Runway 06/24 in Osaka: it's built on a man-made island.
- The highest runway Air Canada lands on: Mexico City, at 7341 ft. (2238 m) above sea level.
- Highest allowable elevation Airbus can land: 12,500 ft. (3810 m) above sea level.

The Deice Man Cometh

As you sit comfortably in your aircraft awaiting a wintertime departure, take a moment to notice the ground crew working outside in freezing temperatures. They may appear to be washing the airplane, but actually they're ridding the wings and fuselage of ice and snow. This procedure is known as deicing and is necessary for your departure, as airline safety standards prohibit takeoff when ice, frost, or snow is adhering to the airplane.

In the early years of commercial air travel, the decision to deice a plane was made by the captain or the airline. Throughout the industry, there was a tendency to resist deicing as much as possible because of time constraints, low operating budgets, and a general lack of knowledge about the perils of ice on an aircraft. Use of technology was limited, particularly for smaller cargo or charter companies whose airplanes sometimes did not have amenities such as heated windshields. In one case, a pilot was equipped with a car windshield scraper to scrape the ice off the plane's windscreen from a side window while on approach.

Although it was technically illegal for an airplane to take off with ice-contaminated wings, a gray area existed because the decision was generally left to the captain's discretion. For example, if a light snow was falling, some pilots would elect not to deice, thinking that the snow would blow off. In most cases, it probably would, but as

The wing gets deiced to help ensure a safe flight

the history books can attest, there are always exceptions.

These days, any second-guessing is removed from the equation, and the old gray area no longer exists. Both Canadian and American regulations now prohibit takeoff when ice, snow, or frost is adhering to any critical surface of the aircraft, including lifting and control surfaces, wings and tail, and upper fuselage surfaces on aircraft with rear-mounted engines. The rule is known as the "clean aircraft concept."

The main exception to the new regulations allows a coating of frost up to one-eighth of an inch thick on lower wing surfaces in areas cold-soaked by fuel, between the forward and aft spars. Deicing also is not mandatory if the captain expects dry snow lying on top of a cold, dry, and otherwise clean wing to blow off during takeoff. For aircraft types where the upper fuselage is a critical surface, a thin coating of frost is permitted in the area provided the deposit is thin enough that underlying surface features such as paint lines, markings, or lettering can be distinguished. Although pilots are in charge of deciding whether deicing is needed, the "lead" ramp attendant can overrule a decision not to deice. Even flight attendants and passengers can voice concerns about the plane's deicing efforts, although the final decision rests with the pilot.

Why a Clean Wing?

Many believe ice on the wings of an airplane is dangerous solely because of the additional weight on the aircraft. However, it is actually loss of lift and the resulting drag on the body of the aircraft that causes problems. Airplanes achieve lift when air flows smoothly over the contoured surface of the wing. If this streamlined flow is disrupted because of ice buildup, decreased lift occurs. A wing can lose 30 percent of lift with just a small accumulation of ice. The stall speed, or the speed at which the wing ceases to be able to keep the aircraft aloft, can decrease by 15 percent with drag potentially increasing by 200 to 500 percent. For example, a unique ice formation composed of clear ice that builds up into a single or double horn on critical surfaces can severely disrupt airflow and increase drag 300 to 500 percent. Meanwhile, ice, frost, and snow that accumulate to the thickness of medium or coarse sandpaper on the leading edge and upper surface of a wing can reduce wing lift by as much as 30 percent and increase drag by 40 percent.

In-Flight Ice Formation

Airframe ice does not occur only on the ground. Although there exist some 30 variables when it comes to the formation of ice on an aircraft in flight, the two primary factors are visible moisture (clouds) and freezing temperatures. Clouds contain super-cooled water droplets, which are composed of water in a liquid state, even though temperatures are below freezing. When a super-cooled droplet strikes an aircraft, it freezes upon impact. To prevent such freezing, airliners are outfitted with heated leading-edge wings that are warmed by the hot air bled from engines. Heated windscreens, instruments, and engine probes and intakes, as well as continuous use of engine igniters, all aid in the battle against ice accumulation.

In turboprop aircraft, electric heaters deice the large rotating propellers. Turboprops also have a rubber cover called a "boot" along the leading edge of the wing. The boot can be expanded during the flight to break off any ice that has attached itself to the aircraft.

Toronto's Central Deicing Facility

No place better illustrates the new era of plane deicing than the Central Deicing Facility (CDF) at Toronto's Lester B. Pearson International Airport. As one of the most northerly countries in the world, Canada must take its plane deicing seriously, and the CDF's

Vestergaard Elephant Betas
armed and ready

massive complex illustrates just how committed the country's airline industry is to safety.

In Canada and similar locales, icing conditions can lurk nearly nine months of the year, so the deicing checklist is always within reach because it's part of doing business. The CDF at Toronto's airport is the largest deicing facility in the world. Fully operational since the 1999–2000 cold season, this 65-acre "drive-through airplane wash" consists of six huge bays capable of handling hundreds of aircraft daily. It has an official deicing season of October 1–April 30. Many pilots jokingly refer to the CDF as the "central delay facility," but the fact that most pilots are paid by the minute takes the sting out of any wait. In addition, the short time it takes to spray a plane with deicing fluid is insignificant compared with the potential for disaster if a pilot did not take the time to deice his or her aircraft. Moreover, the CDF has actually reduced time between deicing and takeoff because it was built closer to the runways and has increased overall throughput and improved turnaround times.

On the way to the CDF, after passengers have boarded the plane, pilots radio "pad control," which assigns the aircraft to a deicing bay. Because this is a "live" or "engines running" operation, precise terminology and electronic signboards are used to eliminate any potential for accidents. Pilots then contact the "Iceman" in the deicing control center, appropriately nicknamed the Icehouse.

Once the aircraft is in position to receive the deicing spray, a machine called the Denmark Vestergaard Elephant Beta springs into action. Smaller planes might need only one Beta for deicing, while larger jumbo jets might need as many as four. The CDF has 27 Beta machines, each of which costs about Can$1 million. The Iceman tells the pilot the exact time deicing started, the type of fluid used, and when the vehicles have retreated to their safety zones. A safety zone is an area ensuring a safe distance between the aircraft and deicing vehicle. The deicing vehicles must be behind these lines before an aircraft can exit the deicing area.

While many airports still employ manually operated "cherry pickers" staffed by ground crew who must brave the bitter winds and back spray, the CDF machines are operated remotely by the Iceman from a heated enclosed cab. They are armed with deicing fluid, nozzles, whisker-like probes to prevent aircraft contact, and a telescopic boom to reach distant spots and critical flight surfaces.

The deicing procedure involves spraying fluids that remove or prevent ice build-up all over the aircraft. Strictly speaking, deicing refers to the removal of existing ice, while anti-icing prevents new ice from forming. Made up of combinations of glycol and water, deicing and anti-icing fluids come in different varieties that each serve a specific function. The difference between the types of fluid is the "holdover time," or the time from when deicing commences to the time the airplane must be airborne, based on temperature, precipitation rate, and type. For example, with Type I fluid at $-3\,°C$ in light snow, the holdover time is about 11 minutes. For most operations, the deicing Type I fluid is used to remove the snow and ice, and Type IV is used to prevent further adhering of ice.

As an airplane is being deiced, all of the extraneous fluid that falls off the aircraft is collected in holding tanks to ensure compliance with environmental regulations, as deicing fluid can be a hazard to nearby bodies of water. The tanks can hold up to 3,434,237 gallons (13 million L) of reclaimed fluid. Some of the spent fluid is used to make car windshield wash and engine coolant, but it cannot be reused for airplane deicing because possible degra-

An Airbus 320 getting a spray in Montreal

dation of the fluid means that its effectiveness cannot be guaranteed. Air Canada prohibits the use of most recycled fluid.

According to Joe Forbes, senior manager of deicing operations at the Greater Toronto Airports Authority, a typical Airbus A320 that holds about 140 passengers in light snow conditions requires 80 gallons (303 L) of Type I fluid and 69 gallons (261 L) of Type IV fluid, with actual deicing time taking just over four minutes. The throughput time at CDF for an Airbus is an amazing 12 minutes.

At more than four dollars per gallon for Type I and double that for Type IV fluid, deicing an airplane is an expensive proposition. During one three-day ice storm in April 2003, the CDF used 396,258 gallons of deicer in a single day, the highest amount in the facility's history. At one point the CDF actually ran out of deicing fluid and scrambled to get more from Chicago, Denver, Forth Worth, and Montreal, Forbes said. One truckload of 4497 gallons that was brought in from Chicago was dispensed on a single jumbo aircraft. Because deicing fluid has a limited shelf life once it has been sprayed on an aircraft, pilots consult onboard charts and consider current temperatures and types of precipitation to determine how long they have before they must get airborne. If the takeoff is delayed for any reason, they may need to head back for a re-spray.

So if your airplane is diverted back to the deicing pad, pass the time by watching for the orange- or green-colored spray of the deicing fluids and the hard-working ground crew braving the blowing snow and wind-chill factor in order to ensure your safety, and the safety of all airline customers.

As we near the runway, ground control hands us off to tower by telling us the tower frequency. Many are surprised to hear that both the ground and tower controllers work side by side in the control tower. These controllers don't miss a beat, handling billions of dollars of aluminum per shift.

Look way up: Toronto's control tower

Don Whitton sits high in his office making split-second decisions without hesitation, millions of dollars in his hands each hour. He's not some high-powered stockbroker, but an air traffic controller sitting 213 feet (65 m) above the ground in the world's most advanced control tower.

Don sits with five others in the cab, tapping touch-sensitive computer screens at Toronto Pearson's control tower. Things have certainly changed in the air traffic control world since the last time I visited a control tower. In my previous visits to other towers, I had walked up a spiral staircase to constant chatter among the controllers. Here, an elevator takes you to the base of the cab, and what is immediately noticeable is how quiet everything is. I'm at the busiest control tower in Canada — a plane lands or departs here every 30 seconds during peak times — and it's as quiet as a library now that verbal communication among the controllers has been replaced by digital communication.

There are normally five operational positions in the control tower, and the staff rotates through these positions during their shift. Operations at the tower are divided geographically with north air and south air towers, and north ground and south ground operations. The fifth position is clearance delivery, where they issue flight plan clearances via voice or data link. During tri-runway operations (the three runways oriented northeast to southwest), two more positions are added: center ground and south tower. A supervisor oversees the entire operation with everything being recorded, including the supervisor's phone conversations.

Presently, Don is working south ground operations. The ground controller tends to be the busiest because of all the aircraft movements to and from the runways, along with the numerous maintenance and service vehicles. Don is one of 40 controllers working at this tower who have come from other towers or control centers across Canada. He's from the Moncton, New Brunswick, control center and is a licensed pilot. He doesn't miss a beat. The information on each airplane comes up on his screen in the form of a flight data strip that provides him with a specific flight number, aircraft type, destination, and other such information.

As the plane makes its way to the runway, Don tells the pilot to contact the south tower controller (who is sitting beside him). The handover is no longer done by paper strips but by a light tap on the computer screen, which will send the information on this flight to the south tower controller's monitor. This system, EXCDS (Extended Computer Display System), was mostly created in-house at NAV CANADA, and it is the envy of many of the countries that have visited the tower to see EXCDS in action.

Don shows me the layout of the airport on another computer screen called Airport Surface Detection Equipment (ASDE). It consists of two radars that are strategically located at the airport to provide a mosaic presentation to ground controllers who can survey all airport movements, making sure aircraft and vehicle operators go where they are supposed to.

After about an hour and a half, Don is scheduled for a break. He can go downstairs to a lounge with large reclining chairs, or even have a short workout in a fully equipped gym at the base of the tower.

During my visit, sunny skies prevailed with a brisk wind from the northwest. Visibility was unlimited, allowing a clear view of the CN Tower and a great appreciation for all the construction that's taking place at the airport. Today they're landing on Runway 33 left and 33 right, two parallel runways that are oriented 330 degrees to magnetic north. It was a pleasure to visit the control tower and put a face to the voice. To see and learn how air traffic controllers move the metal in this high-tech environment certainly was an eye opener.

For a passenger, the unfamiliar aviation sounds may start as early as when you board the plane. The first one may be a hissing sound becoming louder and louder as you walk down the Jetway. It's unsettling when you don't know what the noise is caused by or if it's normal. This particular hissing is the exhaust from the APU. Understanding these aviation noises can help fearful fliers relax and enjoy the flight.

Sounds Good

Ever wonder about the mysterious sounds you hear on an airplane? For all those passengers who've asked me what's behind every thump, whine, and whir heard during a typical flight, here is a rundown of the normal operating sounds that are music to a pilot's ears.

While nestling into your seat, one of the first things you may hear is a whining sound coming from the tail. That's the auxiliary power unit, which provides air conditioning and electrical power to the airplane while it's on the ground. We usually activate it about 10 minutes prior to pushback because the air from the unit is used to start the jet engines. You may also hear a sound as the lights in the cabin momentarily flicker. This occurs when the external power is removed and the onboard electrical system takes over. The hissing noise that follows is the air flowing through the gaspers, or air conditioning ducts, above you.

Meanwhile, from below you'll hear the thump, thump, thump of baggage being loaded. There may also be a high-pitched hum from the electric hydraulic pumps that help operate the cargo doors. A loud clump signals the cargo doors or main cabin door closing, indicating it's time for pushback.

During pushback and engine start, the cabin will become very quiet as the airflow is channeled into starting the engines. After the engines are started you will hear them whining as the pilot increases the power to taxi the airplane. Then comes the electric-motor sound of the display screens descending in unison to brief you on the airplane's safety features — a sound you will hear again as the screens stow for takeoff.

During taxi there may be creaking noises as the airplane wheels go over the uneven taxiway. Like roads, not all taxiways are perfectly smooth. As the pilot taxis to the runway, the flight crew makes the one-minute-before-takeoff announcement. You will hear the engines roar when the power is increased to allow for takeoff. Today's efficient engines are much quieter than those of yesteryear, so the cabin is quieter as well. Sometimes there may be a thumpity, thumpity, thumpity noise as the airplane rolls down the runway. This is the nose wheel going over lights embedded in the centerline. It's now takeoff time.

Airbus 320 thrust levers

I Have Control

Once we pilots finish our before-takeoff checks, we make a flight-attendants-take-positions-for-takeoff announcement. The pilots take a quick look to ensure no one is on immediate final approach, and we turn on lights to make our presence known to other pilots.

Finally, I read "Cleared for takeoff" back to the control tower and the big takeoff lights are turned on, signifying all the checks are done — we're going flying! The captain hands me the control of the airplane and says, "You have control," which I acknowledge by saying "I have control."

About 65 departments within the airline and airport have taken us to this point.

One way to describe the job of an airline pilot: moments of shear exhilaration coupled with hours of boredom.

---

56 Seconds Barreling Down the Runway

Now that we are cleared for takeoff, I position the Airbus A340 on to the centerline of the runway, smoothly advancing the thrust levers to their takeoff detent (their takeoff position). In other airplanes the thrust levers are also called the throttles. Similar to an accelerator in a car, the thrust levers control the power to the engines. I call out a few instrument indications as the four engines spool up, forcing everyone to lean back into their seats. Even after thousands of flight hours, the sound of the thrust still exhilarates me.

My right hand leaves the tiller, a device that steers the plane on the ground via the nose wheel, and grasps the joystick, which is forward and to the right of my seat. My right wrist sits on a very expensive armrest that was precisely adjusted during our preflight checks. A smooth rotation of the wrist is all that is required to get this mammoth bird airborne. The fly-by-wire joystick, similar to the one used with computer games, is one of the many things that make the Airbus a unique aircraft. This state-of-the-art technology comes with a price tag of Can$200 million for each plane.

After setting the takeoff power, my left hand leaves the thrust levers and the captain's right hand takes over in the event we would have to abort. Protocol states that only the captain can perform a rejected takeoff. In that event, he or she would move the levers to the maximum reverse position.

My heart rate increases a few beats and my concentration is at its highest. I am focusing down the runway, ensuring the airplane is on the centerline, using foot pedals to move the rudder, located on the tail. Meanwhile, the captain scrutinizes the engine instruments. As the speed and momentum builds, this airliner (one of the world's largest) only requires small rudder inputs to keep it straight as it rolls down the runway.

Our ears are in tune for anything irregular, but we only hear a symphony of nearly 212,000 pounds of in-sync engine thrust. We feel repetitive bumps as the nose wheel rolls over the lights embedded in the runway.

We are departing from Runway 23 at Toronto's Pearson International Airport, oriented 230 degrees to magnetic north. There is a stiff wind from my right, forcing me to work a little harder to keep the plane straight, but it's well within the limits of this craft. The captain calls "100," indicating the airspeed is 100 knots (115 mph or 185 km/h), to which I respond, "Roger." This mandatory call serves as an airspeed cross-check — ensuring our computer-driven instruments are up and running — and a pilot-incapacitation check.

While I keep the aircraft on the runway centerline, the captain concentrates on scanning the high-tech glass instruments, now paying particular attention to the building airspeed. The captain decisively calls "V1," which is a pre-calculated speed based on weight, wind, temperature, runway, and other factors; it is similar to a yellow traffic light. In other words, it is decision time: Do we stop? Do we continue?

If everything is normal, the captain removes his or her hands from the thrust levers, which means we're continuing with the takeoff! Finally, a crisp "rotate!" is heard from the captain, based on another pre-calculated speed. I take a deep breath and silently count a slow one, two, three, supplying gradual backward pressure on the joystick to achieve flight. At first it seems nothing is happening, but the 208-foot (64 m) wingspan has already begun to lift. On

An A330 takes flight

an interesting note, the distance of the Wright Brothers' first flight was shorter than the width of this plane's wingspan.

As the wings bite into the air, the aircraft's nose smoothly lifts and a small clunk is heard as the main wheels leave the runway. This runway, the longest at Pearson, seems pretty short at this point.

More gradual backward pressure is applied to the joystick and the captain calls "positive rate," indicating the airplane is climbing and it is time to bring up the landing gear. You do not want 18 metric tons (20 tons) of landing gear hanging out in the wind, so I command for the gear to go up by saying, "Gear up!" Another light thump is heard as the gear tucks into the belly of the aircraft. Today's takeoff took an exhilarating 56 seconds.

I then set the climb power and the captain checks in with Toronto air traffic control. This will be the last time I touch the thrust levers until seconds before touchdown. The Airbus' advanced auto-thrust system calculates and sets its own power settings, much like cruise control in a car. The autopilot is engaged and it's time to unwind. It's hard to imagine the landing is 16 hours away.

A Typical Flight

As you board Air Canada Flight 856 en route to London from Toronto, you quickly glance into the flight deck of the Airbus A340 and wonder what exactly goes on up there before and during flight.

Well, we the pilots have been on board for about 45 minutes. There is a lot to do. There are checks; the flight plan has to be entered into the navigation system, radios must be monitored and clearances obtained. Everything is performed by standard operating procedures. That means when a pilot flies with another pilot, he or she can expect the same procedure and response every time.

Settling into the cockpit, we get out all our charts, buckle up our seat belts, don headsets to communicate with air traffic control (ATC), and receive instructions during each stage of flight. At many airports, the Automated Terminal Information System (ATIS) is now available through a data link to the flight deck, which cuts down on radio communications considerably. (Even our air traffic control clearance is now sent to the flight deck via data link at many airports.) ATIS provides information on runways, winds, temperature, and other factors. We use these details, along with data on aircraft weight, airport elevation, runway length, and icing conditions, to determine the ideal takeoff speed and whether the runway is suitable for the aircraft's weight.

After all the cockpit checks are completed, the captain and I review specific emergency procedures to spell out our individual responsibilities. This is a requirement for the first leg (flight) of the pairing. It is not done for the remaining flights providing the flight crew stays the same. To keep an ongoing record of everything that goes on, all flight deck conversations are recorded by the cockpit voice recorder.

A vast number of people both inside and outside the aircraft work hard to get this state-of-the-art airplane airborne. There are ground personnel doing everything from loading baggage and emptying the lavatories, to adding water to the two large onboard reservoirs. The fueler is uplifting the flight plan's fuel. For today's flight, 45 metric tons are added to the 13 already in the tanks for the required total of 58 metric tons.

On board, groomers, commissary, and flight attendants work diligently. Maintenance is also present dealing with any snags on the aircraft. (Some common snags, for example: seats not reclining, coffee makers not working, clogged sinks, and temperamental video systems.)

Approximately 15 minutes prior to pushback, the ramp lead plugs his or her headset in near the nose wheel and gets a commu-

The flight deck of the A340

nication check from the cockpit. The captain responds by saying "Good day, communications OK, brakes set." In the cockpit, we have completed our checks and now respond to requests from various staff. One flight attendant is requesting a cooler cabin temperature so we start the auxiliary power unit (APU). This is really a fifth jet engine located in the tail to supply air conditioning and backup power. The passengers started boarding 30 minutes prior to departure, but there are two passengers just going through security now. An airport agent tells us to expect a five-minute delay. Minutes later, dispatch informs us the two passengers have arrived and we are ready to go. The service director states that the cabin is secure. That is, everyone is in his or her seat.

A pushback clearance is received from the ramp or apron controller and the ramp lead is told "cleared to push, brakes are off." The aircraft moves back, one minute behind schedule. The aircraft automatically sends this updated departure information to company operations.

The before-start, after-start, and taxi checks are all performed prior to takeoff. The service director enters the flight deck to confirm a full load of 284 passengers. A welcome-aboard announcement is made after the safety demos in the cabin are completed.

We are number four for departure off Runway 23 and watch as the plane ahead of us takes off. We are cleared to position and the announcement, "Ladies and gentlemen, we are taking off in one minute. Flight attendants take positions for takeoff" is made. We acknowledge our takeoff clearance, turn on the landing lights, and progressively apply takeoff power. Fifty-two seconds later one of the most sophisticated airliners in the world is airborne. Once the landing gear is retracted, the autopilot is engaged and it makes a gentle left turn as per the clearance.

Departure control is contacted, and we are vectored on course and cleared to higher altitudes. We are passed on to Montreal control, who then clears us to flight level 330 (33,000 feet or 10,058 meters above sea level). Each area we fly over has a different set of controllers. Tonight we will talk to Toronto, Montreal, Boston, Moncton, and Gander air traffic control centers on this side of the "pond." The aircraft is light enough to climb to our optimum altitude of flight level 370 (11,278 m). Aircraft are seen passing over, under, and across us throughout the evening as we are vectored to fit into the designated slots for the Atlantic crossing.

The Airbus A340 is flown by two pilots; however, three and even four pilots may be required depending on the length of flight. The

The crew bunk in an A340

extra pilot(s) share in the workload allowing the others to rest up in a bunk directly behind the flight deck or in a business-class seat. Everyone is in the flight deck for takeoff and descent.

Thirty minutes prior to landing, a descent is initiated as calculated by the onboard computers. London control vectors us for the instrument landing system (ILS) to Runway 27R. The wheels touch down 10 minutes early because of stronger than normal tailwinds.

There to meet us are the ground staff directing us by hand

signals waved with florescent wands. As well, a lighted guide-in system aids to precisely position the aircraft at the gate. Another crew; a clean, restocked, and refueled aircraft; and new passengers start their journey a few hours later.

CHAPTER FOUR

After-Takeoff
and Cruise Checklists

Flight, Weather, and Turbulence

As the landing gear tucks into the Hong Kong–bound aircraft's belly and it picks up speed and altitude, the captain checks in with Toronto departure, one of the many controllers we will talk to today. He tells the controller our call sign, "Air Canada 015 heavy," our altitude passing, altitude cleared to, and which runway we departed. We mention the word "heavy" not because we are near our maximum weight, but because it is protocol for every airplane heavier than 300,000 pounds (136,000 kg), at least here in North America. (A heavy aircraft must be given a wider birth from smaller aircraft by air traffic control because of potential turbulence emanating from the wingtips.)

The flight deck is sterile below 10,000 feet (3048 m), meaning idle chitchat is kept to a minimum. The seat belt sign remains on, and if smooth conditions exist it will be turned off, along with the bright takeoff lights, when passing through 10,000 feet. It will take six hours to reach the North Pole with many of our position reports passed on to Arctic Radio.

Most would assume Arctic Radio is strategically nestled at an Arctic airport. It was established 600 feet (183 m) underground, located in NORAD's (North American Aerospace Defense Command) base, in North Bay, Ontario. I called the facility and Ted Graham, flight service specialist, informed me they moved to the surface in September 2006. There, seven specialists cover the entire

Arctic (Alaska to Yellowknife in the west, and Baffin Island and Labrador to the east). Some of their duties include filing flight plans, disseminating weather reports, and noting and recording aircraft positions. He says he doesn't miss the over-a-mile-long bus ride down to the site; however, the new military site still doesn't have any windows.

Controller: You have traffic at 10 o'clock and 6 miles!
Pilot: Give us another hint, we have digital watches!

While conversing with the next sector's controller, he gives us a heads-up on some light turbulence as we level off at our cruising flight level. Air traffic control is a great means to relay the where-abouts and intensity of turbulence between pilots.

Pilots always try to deliver a smooth flight to passengers, and a safe work environment for flight attendants. And with the seat belt sign on we don't get beverage service either, so a much-needed coffee may be a long time coming. Because we fly through a medium always on the move, both vertically and horizontally, it's inevitable we'll run into ripples and bumps. So why the bumpy ride? And what can pilots do about it?

Why the Bumps?

The bumps sometimes experienced when flying are not caused by illusive air pockets, as is commonly believed. Sudden in-flight movements are caused by disturbed conditions in the atmosphere, which are called irregular wind currents, or turbulence. There are six natural types of turbulence, which occur under different conditions during the phases of flight. Fortunately, your flight crew is prepared to avoid all six, thus ensuring your comfort.

Mechanical turbulence occurs when an aircraft encounters strong winds blowing over irregular terrain such as hills, trees, or buildings. This type occurs near ground, at less than 4000 feet (1220 m). To avoid mechanical turbulence after takeoff, the pilot will steepen the angle of the aircraft's climb. When landing, he or she will decrease the aircraft's speed.

The second kind is known as convective turbulence. During the day, the sun heats up the Earth, which then heats the overlying air. The hotter the air, the bumpier the turbulence. Convection, the process by which heated air travels upward in the atmosphere, is at its maximum during the heat of a summer afternoon. Cloud formations are good indicators of the degree of convection at work, and if a pilot sees white, puffy, convective, or cumulus clouds, it is often a sign of turbulence. He or she can then easily avoid it by flying above the clouds.

When these billowy clouds become taller than they are wide, they are called towering cumulus and potentially signal the buildup of thunderstorms. Once these clouds are moderately or fully developed, they show up on the aircraft's radar — not the clouds themselves, but the rain showers inside. The larger and more intense the rain shower, the more obvious they appear on the radar. Pilots can then avoid the turbulence by navigating around the showers.

A third type of turbulence is low-level wind shear (LLWS). It is the result of rapid wind shifts between updrafts and downdrafts under a thunderstorm or winds funneling down a valley. Pilots are familiar with this hazard and therefore avoid flying under thunderstorms, particularly during takeoff or landing. Many aircraft now have wind-shear detecting equipment on board, and a number of airports have equipment to detect LLWS.

Orographic turbulence occurs when a strong wind blows perpendicular to a mountain range, causing a phenomenon known as a mountain wave (orography is the study of mountains). By watching for the three types of clouds that may indicate its presence, a pilot can either avoid the bumpier air of a mountain range's downwind side or climb above it.

A fifth type, called frontal turbulence, is brought on by a sudden change in wind direction due to a weather front. A quick look at the latest weather chart will tell a pilot where these fronts are, as well as the speed at which they are moving.

The last, and perhaps least understood kind of turbulence, is clear air turbulence (CAT). This forms when little to no weather systems are present (hence the name clear air) and occurs most often near jet streams — long, thin bands of fast-moving air sandwiched between the first two layers of the atmosphere, usually located at 36,000 feet (10,973 m). Jet streams corkscrew around the globe like coiling, meandering snakes, and they occur when two air masses collide (see below). The rapid increase in wind speeds (wind shear) can cause significant turbulence, but can usually be avoided without difficulty.

A seventh type of turbulence exists but it's man-made, coming directly from another airplane. Lift is the result of a high-pressure area below the wing pushing the wing against a low-pressure area on top of the wing. The resultant movement of air from the bottom to the top swirls rapidly around the wingtips and quickly outward, which can make for a very bumpy ride. We name this type wingtip vortices or wake turbulence. We know wake turbulence moves outward and downward, so if we stay above the preceding aircraft's path, or far enough away, we can avoid these man-made bumps. You'll notice many modern airliners have vertical wings attached to the wing tips. Called winglets, their job is to reduce this swirl, thus minimizing induced drag and improving overall performance.

Today's onboard navigation devices, enhanced by satellites, are incredibly accurate, so it's not unusual to find a plane directly below or above in places like the Atlantic Ocean, where the roads we travel are few. It's a normal occurrence to see a cluster of airplanes at various altitudes en route to and from Europe over the Atlantic. (This just confirms the fact we are not lost.) A simple offset of just one mile is usually enough to steer clear of the bumps from an airplane above us.

The intensity or strength of turbulence is very subjective; one person's "rough ride" may be just a "little bumpy" to another. It's hard to pinpoint, comes and goes, and also depends on the size of aircraft. Various intensities of turbulence exist but rest assured the members of your flight crew are always up-to-date on the latest weather reports and forecasts, and they are expertly trained to avoid, or at least minimize, any discomfort. So sit back and relax; the next bump you experience will be your landing!

Sky Snakes

We all probably have some vague notion of what a jet stream is. Those streaky white trails that appear behind high-flying jetliners? Well, no, that's a common mistake. Those are condensation trails formed by the exhaust of aircraft engines (see sidebar page 86). Jet streams are actually a much more spectacular phenomenon — long, thin bands of extremely fast-moving air that form at high altitudes and corkscrew through the atmosphere around our planet.

In 400 BC, Aristotle wrote a treatise on weather entitled *Meteorologica*, in which he noted that higher clouds may move faster than lower clouds. Over 2000 years later, when manned balloons were first launched during the 18th century, those aboard noticed that winds tended to increase with height. But it wasn't until World War II and the advent of high-altitude flight that jet streams were encountered and their presence confirmed.

On November 24, 1944, 111 U.S. Air Force B-29 bombers were sent from the Pacific island of Saipan to attack industrial sites near Tokyo, Japan, in the first high-altitude bombing mission of World War II. As the airplanes approached the island of Honshu at 33,000 feet (10,058 m), they were suddenly hit by winds of 140 knots (161 mph or 260 km/h), which knocked them completely off course. Only 16 of the 111 pilots managed to hit their targets, while the rest were blown over the ocean and forced to return to base. Pilots on subsequent missions also reported encountering extremely powerful winds and unexpected turbulence when flying westward to Japan. What were these incredibly strong winds? One pilot likened them to a jet of air streaming out of a hose with enormous velocity; hence the name jet streams.

We now know that jet streams are produced as a result of the significant temperature changes where air masses collide. In North America, for instance, where there may be four distinctive air masses, up to three separate jet streams would exist. They can be hundreds of miles long, tens of miles wide and a few thousand feet thick. (I liken their dimensions to a Christmas ribbon when teaching my weather classes.) Migrating southward in winter and decreasing in altitude the further north they are found, jet streams are strongest during the winter months because the frontal zones or temperature differences between air masses are more dramatic in winter. They mark the dividing line between seasonable and unseasonable temperatures. They also indicate in what direction and at what speed surface weather (highs, lows, fronts, etc.) is traveling.

To locate jet streams, weather balloons are sent up to penetrate the higher atmosphere, climbing to 100,000 feet (30,480 m). As well, airliners have equipment onboard to gauge upper atmospheric wind conditions, and satellites capture some of the telltale cloud patterns associated with jet streams.

Sometimes a pilot will want to get into this fast-moving air; at other times he or she will want to avoid it. We try to capitalize on strong tailwinds and, if able, duck out of the stronger-than-normal headwinds.

If any turbulence is detected, pilots always ask for ride reports from air traffic control and then climb or descend to find smoother air. On occasion, this air can also be rough and become so quite quickly, which is why regulations have you keep your seat belt fastened at all times.

Streaking White Lines

An undulating contrail indicates turbulence

➣ Name of white streaks in sky left by airplanes: contrails (short for condensation trails) or vapor trails.

➣ Reason for these white streaks: water vapor in jet exhaust freezes upon impact with very cold air.

➣ What a formation of contrails resembles: seeing your breath when you exhale into cold air.

➣ Main composition of contrails: ice crystals.

➣ Typical temperature at which contrails start to form: −40°C (−40°F).

➣ Typical altitude at which contrails start to form: 28,000 ft. (8534 m).

➣ Typical cruising altitude of jet airliners: 37,000 ft. (11,278 m).

➣ Duration of contrails: several seconds to hours.

➣ Factors that contribute to long duration: moist air and/or light winds.

➣ What long-lasting (or persistent) contrails may evolve into: cirrus clouds.

➣ Factors that contribute to quick evaporation of contrails: dry air and/or strong winds.

➣ Typical dimensions of contrails: tens to hundreds of miles in length, several miles in width, and 300–1300 ft. (90–395 m) in thickness.

➣ Estimated contrail coverage over the Earth: 0.1%.

➣ Regions with highest percentages of contrail coverage: U.S. and Europe.

➣ Significance of a thick, widening, long-lasting contrail: humid air high in

the atmosphere, a possible early storm warning.
➤ Significance of twisting contrails: turbulence.
➤ Event that allowed research into skies without contrails: all flights in U.S.
grounded after September 11, 2001, attacks.

Weather Watch

Your airplane may be doubling as a flying weather station because
of a new system, the Aircraft Meteorological Data Relay (AMDAR),
which disseminates much-needed weather data. Most aircraft continually gather and transmit some 400 pieces of data to the ground
through data link, and some of that data is meteorological. Because
of this new access to information that is processed by weather
supercomputers, pilots and meteorologists have benefited from
better predictions of high winds, precipitation, cloud heights, and
turbulence, among other phenomena. Your airplane may be helping to forecast weather for everyone, as well as helping to bring you
even smoother flying in the years to come.

For more than 50 years, upper air observations have been made
with other technology — radiosondes. These are small, expendable
instrument packages suspended below a large balloon filled with
hydrogen or helium.

As the radiosonde is carried aloft, the battery-powered sensors
measure pressure, temperature, and relative humidity. All of the
information collected is relayed to sensitive ground receivers. A
radiosonde flight can last in excess of two hours, and during this
time it can rise to over 100,000 feet (30.5 km) and drift more than
125 miles (200 km).

When the balloon expands beyond its elastic limit, it bursts and
a small parachute deploys to slow the descent of the instrument
package. (What goes up must come down!) Worldwide there are
nearly 1000 upper-air observation stations and each station releases
a radiosonde at 0000 UTC and 1200 UTC every day of the year.
Through international agreements, radiosonde data is exchanged
between countries. It is interesting to note that of the 75,000
radiosondes released by American stations each year, only 20 percent are ever found and returned for reconditioning and reuse.

As the balloons ascend at a rate of approximately 1000 feet/

minute (305 m/min.), they are tracked to determine the exact wind speed and direction. Radiosonde observations are applied to a broad spectrum of efforts. As with the aircraft weather data, these efforts include data initialization for global and regional computer-based weather-prediction models, input for air pollution/dispersion models, severe storm modeling, verification for satellite data, weather research, and climate change studies.

Now that you've boned up on jet streams and the intricacies of vapor trails, let's discuss how winds impact flights. Whether the wind is blowing at the surface or zipping along at high altitudes, a pilot must contend with it.

All modern airliners have wind read out in the flight deck. This same wind info is included on the moving map displays found projected onto the cabin screen or on the individual entertainment displays in many airliners. From accurate navigation equipment one can determine where and how strong the winds are outside.

Recently, while exiting the flight deck for physiological reasons (people are surprised to see pilots out of the flight deck using the washroom), I noticed a passenger intensely watching the overhead screen projecting our progress on a map. I jokingly said, "Ah, that's where we are!" His look changed from total shock and horror to a wide grin and laughter.

Wind Beneath the Wings

"Ladies and gentlemen, due to strong tailwinds we will be arriving at our destination 20 minutes early." When passengers hear an announcement like this, they are often surprised. But the wind outside the airplane window is one of the most important weather parameters for a pilot and a powerful force affecting the aircraft.

The wind at flight altitude can be well over double hurricane strength — 150 knots (173 mph or 278 km/h) is not unusual. When the wind blows in the same direction in which the aircraft is flying it's called a tailwind, which can greatly increase the aircraft's speed.

Conversely, an airplane heading directly into the wind, called a headwind, has to work against the current, which slows it down. But what causes this movement of air? There are two main factors involved: the Earth's spherical shape and its rotation.

Because of the Earth's shape, the sun's rays are more intense at the equator than at the two poles. Thus different parts of the Earth heat up at different rates, which in turn causes unequal heating of the air. These temperature variations cause differences in air pressure. Those Hs and Ls seen on weather maps refer to systems of high pressure and low pressure, which are what set the air in motion, thereby creating wind.

The Earth's rotation causes a phenomenon called the Coriolis force. It makes everything that is in motion in the northern hemisphere, including air, deflect to the right. Wind is therefore seen to blow counterclockwise around a low-pressure system and clockwise around a high pressure system. The same phenomenon occurs in the southern hemisphere, only everything is reversed.

In most of North America, the prevailing winds are the westerlies, which move from west to east. (Winds are named after the direction they're coming from, not the direction they're going in.) If you were to peruse a flight itinerary, you would notice the flight from Vancouver to Toronto is 28 minutes shorter than the return flight. If your travels take you to the southern Caribbean, you may discover that the prevailing winds there, the trade winds, are generally from the east. Local phenomena such as sea breezes and daytime heating further complicate wind patterns.

As you may recall, winds generally increase with altitude and tend to be weaker during the summer. In winter, winds at flight level can exceed 200 knots (230 mph or 370 km/h). A pilot's weather briefing always includes the whereabouts of jet streams, which can cause turbulence and headwinds that result in a longer than usual flight. Pilots, along with flight dispatchers, have access to weather charts depicting these "sky snakes," whose location and intensity are constantly monitored and updated.

Winds also play an important role in takeoffs and landings. Because of the aerodynamics involved, pilots prefer to take off and land into the wind, but sometimes the winds do not cooperate. Pilots must frequently deal with winds that blow across the runway, called crosswinds. When watching an airplane land, you may notice that its body does not seem to be aligned properly with the runway on the approach, and that its nose appears to be off center. The airplane may cock into the wind — a sideways movement called crabbing. The pilot is in fact positioning the airplane slightly upwind and using a combination of rudder and aileron to compensate for the crosswinds, which are pushing the airplane sideways. While a landing where the upwind wheel of the landing gear touches down first, with one wing lower into the wind, may look poor, this is not indeed the case. It is typical of a crosswind landing, which, with a bit of finesse, can be very smooth. The key, as in any aspect of aviation, is understanding and respecting the elemental force of the wind.

When a pilot makes cabin announcements, certain words should be avoided. For example, a pilot explaining that strong headwinds may cause a delay of 20 minutes might be tempted to say it's "due to bucking headwinds." Once through the PA system, however, the word "bucking" may sound like something else to the passengers. . . .

As we level off at 34,000 feet (10,363 m), our initial cruising level, meal service is well underway. I'm sure you won't be surprised to hear your gastronomic experience is the equivalent of picnicking on a mountain several thousand feet high. You may even find that you become a little bloated, your feet swell, or your plastic water bottle expands a little. That's because you're feeling and noticing the effects of pressure lower than what you experience on the ground. This is something to consider when you have an alcoholic drink in flight; it has a greater effect on you at higher heights because of the decreased oxygen. Ever notice your coffee doesn't seem quite as hot in flight? That's because it isn't! Water boils at about 94°C (201°F) instead of 100°C (212°F). Food is chosen by airlines with the consideration that smells are less pungent while dining in the Cabin Alps.

Under Pressure

We live at the bottom of an invisible ocean of air called the atmosphere. Although you can't see it, this ocean has many properties that must be taken into account when building and operating passenger aircraft. Not only is the atmosphere compressible and expandable, but it has immense weight and is marked by rapid changes in temperature, air pressure, and density.

The higher you go in the atmosphere the colder it gets because it is the Earth, warmed by the sun, that indirectly heats the air. At a cruising altitude of 37,000 feet (11,278 m) a typical outside temperature would be –57°C (–71°F). What's more, the higher you fly, the less air pressure there is. In fact, when a passenger aircraft is at cruising altitudes, the outside air pressure is only about 20 percent of that at sea level.

A third atmospheric property, density, also decreases with height. Because the air is very thin at high altitudes, oxygen is extremely sparse. So in order to fly at these heights, the air inside an aircraft must be put under artificial pressure. Even then, pressurization does not create a cabin environment with sea-level pressure conditions. In fact, at cruising altitudes, the air density is equal to that found atop a mountain 5000 to 8000 feet (1524–2438 m) above sea level.

When the air is less dense, aircraft encounter less resistance, making it easier for them to go faster. Higher altitudes also allow aircraft to soar over most weather disturbances and to increase their fuel efficiency. It is no wonder that with the onset of passenger travel, technology was improved so that airplanes could fly ever higher.

Modernization in the early '40s saw the introduction of pressurized cabins with a more streamlined, cylindrical design. To this day, aircraft fuselages have retained this shape, which ensures a uniform level of pressurization. Passenger cabins must be airtight, so outside air is ducted from one of the compression chambers of the jet engine. From there, this so-called bleed air is cooled by air-cycle machines and heat exchangers. The now room-temperature air is pumped into the cabin and cargo holds. To have a continuous flow of fresh air, some air is then vented to the outside by a set of outflow valves, and the remainder is recirculated.

The overall pressurization system is designed to raise and lower pressure values within a moderate range for passenger comfort. As soon as the aircraft takes off, the system goes to work and starts pressurizing. Likewise, when a plane starts to descend, the pressurization computers automatically start depressurizing the cabin. This rate is far slower than that of the actual climb and descent of the airplane. For example, a cabin typically depressurizes at a rate of 300 feet per minute (91 m/min.), whereas the airplane itself descends at 1500 to 2000 feet per minute (457–610 m/min.). By touchdown the cabin pressure is the same as that outside. The change in pressure may create some discomfort, particularly for those with a head cold and for small children.

During ascents and descents, air must be replenished through the Eustachian tube in the middle ear cavity to equalize pressure with the atmosphere. When air is trapped in the middle ear, it is painful. During ascent, air escapes through the tube easily. Unfortunately, during descents, when pressure in the middle ear must be increased, the Eustachian tube does not open so readily. The situation is aggravated with a cold, allergies, or an infected throat.

To minimize discomfort you should make a conscious effort to swallow on descent, which will help equalize the pressure in your middle ear with the pressure of the atmosphere. Sucking on a lozenge can help, as can offering little ones something to suck on, like a bottle, just prior to descent. Also, remember that it's very dry· outside air that is used in pressurization, and as a result the relative humidity in the cabin tends to be low — around 15 percent. The air has about the same humidity as the desert air you find in Palm Springs, California. Be sure to replenish your body with fluids, particularly on long flights, but stay away from diuretics like coffee and alcohol, which cause fluid loss.

One final word of advice: make sure that the tops to your toothpaste tubes and shampoo bottles are securely tightened; pressurization has been known to cause the odd lid or top to pop open (or pen to explode). This even happens to seasoned aircrew. Taking these simple precautions can help you keep the pressure at bay the next time you're in flight.

A Breath of Air

On a clear day, an airplane flying high overhead looks like a small speck surrounded by nothing but sky, and it's tempting to think of the air it's flying through as a huge expanse of emptiness. But that is far from the truth. Air is a fascinating medium with unique properties that play a vital role in aviation — both inside and outside the airplane.

Four invisible layers of air surround the Earth, creating the

Cloud development in the troposhere

planet's atmosphere. The layer we live in, and where most weather occurs, is called the troposphere. Jet engines are most fuel-efficient at higher altitudes, so we tend to fly near the top of the troposphere or the bottom of the next layer, the stratosphere.

The boundary between these two layers, located at an altitude of around 36,000 feet (10,973 m), is called the tropopause. This is where jet streams, those fast-moving wind currents, may be lurking. It's important for pilots to know the height of the tropopause; it appears on their flight plans and weather charts. The top of the troposphere is where the anvil shape of a thundercloud forms. The flattened shape of the cloud is caused by the tropopause. The third layer, the one above the stratosphere, is called the mesosphere, and the fourth is the thermosphere.

The Earth's average air temperature of 15°C (59°F) decreases by about two degrees Celsius (4°F) for every 1000 feet (305 m) you rise above sea level. So at 36,000 feet, where the stratosphere begins, the temperature is an icy –57°C (–71°F). Atmospheric conditions also vary depending on the day, time, season, and latitude. To keep on top of the fluctuations, weather balloons are launched twice a day from hundreds of locations around the world (see page 87).

Air consists mostly of oxygen and nitrogen, but one to four percent is water vapor. For such a small percentage, the water in the air carries a lot of weight. It is the main component responsible for our weather, and makes the clouds, rain, snow, and fog, which are all very significant factors in aviation. Clouds are formed when air expands and cools (condensing the water vapor in the air). Conversely, when air is compressed and sinks under the influence of high pressure, it heats, causing clouds to dissipate. And so, watching clouds can tell a pilot what sort of flying conditions will prevail.

Warm air has less density than cold air, and it therefore rises above the colder air. This is important to a pilot because it factors into an airplane's stability as it passes through the sky. When warm air rises, say from the heating of the land during the day, colder, denser air will descend to replace that warm air, creating air currents. Such currents cause the air to become less stable and may create bumpy flying conditions. Stable air, of course, is a prerequisite for smooth flying. (See turbulence, page 82.)

Humidity is another factor that determines air density — the greater the moisture in the air, the lower the air's density. A student pilot will learn that the best conditions for flight are found on a cold, dry winter day; a propeller is said to have more "bite," and the denser air enhances aircraft performance.

The atmosphere is compressible and expandable, but it has immense weight and is marked by rapid changes in temperature, air pressure, and density.

Flying north still en route to Hong Kong, the radio chatter diminishes and the traffic in the skies thins out as we fly over sparsely settled northern Canada. But just because we're nearly alone up here, it doesn't mean we can deviate from our path. We must stick to the script, and if a change in path is requested, maybe due to weather, we will be certain to communicate that information.

Rules of the Road

Being a good pilot, like being a good driver, is more than just mastering the controls; it also means abiding by the rules. In aviation, however, the rules and regulations are far more numerous and complicated; they are introduced early in training, and pilots never stop studying to keep abreast of them.

In air travel, noise regulations are a fact of life, just as they are on city streets. During early stages of training, pilots learn that they can't fly low over fur and poultry farms because the noise can agitate the animals and incite them to stampede. The days of barnstorming, low-flying thrill rides through the countryside, are long gone. Flying low to take aerial photographs over penitentiaries is also forbidden. And to lessen the impact of noise on local residents, more and more airports are setting curfews for takeoffs and landings and establishing mandatory altitude levels. Toronto's Pearson International Airport requires that airplanes climb to 3600 feet (1097 m) above sea level before proceeding on course.

Once on course, commercial airlines follow air routes that are like highways in the sky. Separate lanes are created at different altitudes according to the direction the airplane is traveling. Pilots must stay strictly on these routes. Commercial aircraft send signals to the air traffic control (ATC) radar, and a deviation of more than 100 feet (30 m) will automatically be queried.

Even though there are no signs in the sky posting speed limits, pilots must follow a strict set of regulations related to airspeed. Airplanes flying below 10,000 feet (3048 m) must slow to 250 knots (288 mph or 463 km/h). Aircraft must slow even more as they near the airport and start to descend. Even at cruising altitude, airspeed can change by only five percent, otherwise ATC must be advised. Pilots don't get speeding tickets, but they do receive a letter for any infraction and, yes, licenses can be revoked for repeated offenses.

Rules like speed limits can be especially challenging when flying internationally, because they vary from one country to another. Just like North Americans driving in Britain must stick to the left side of the road, pilots need to be aware of all the international differences in air traffic regulations. For example, holding patterns are different for Canada, the U.S., and Britain, so pilots need to know how fast they can go at different altitudes, and how long they should spend in the holding pattern.

A plane's place of origin is always identifiable by that series of large registration letters on or near the tail (see page 35). Pilots of smaller aircraft use these letters to identify themselves to ATC. Further identification is provided by an extensive set of papers that must be carried on board airplanes at all times, including registration, proof of ownership, certificate of airworthiness, and pilot licenses. In Canada, government officers from Transport Canada can inspect these papers at any time, just like a police officer can ask for your driver's license during a roadside inspection.

Weather conditions vary greatly, so there are rules for each set of conditions. Pilots have a number of weather considerations and a number of weather-related questions they need to ask themselves while in flight: Do I have good enough visibility to legally land or take off? Do I have enough fuel, and enough contingency fuel, to fly from A to B? Is there any ice or snow adhering to critical surfaces like wings, fuselage, and flaps? Airline pilots are constantly watching the weather, before and during the flight, to make sure they are following the right procedures. On occasion, they need to consult the rule books because some regulations are very specific and complicated. For this purpose, the flight deck has an onboard library, and many pilots also carry additional manuals with them.

Pilots have a dizzying assortment of rules to follow each step of the way, from flight plans to speed limits to landing protocols. Nowadays it is nearly impossible to just jump into an airplane and fly.

Highways in the Sky

Anyone who spends any time in airports knows that air traffic is on the increase. And as the highways in the skies fill with aircraft, air traffic control systems are becoming increasingly important and sophisticated. Canada, which has one of the safest air traffic control systems in the world, can have hundreds of commercial airplanes in its airspace at any given time. Keeping track of them all is a vast cross-country network of air traffic controllers, flight service stations, and navigational devices.

Equipped with a radar screen that depicts a particular area, air traffic controllers transmit their clearances and requests for information to aircraft via radio. They use terms standardized by the International Civil Aviation Organization (ICAO) in order to avoid any potential misunderstanding. Each part of the aircraft's journey is the responsibility of a different controller. For example, the pilot of a flight from Toronto to Montreal — the busiest route in Canada — will communicate with 15 different controllers by the time the passengers disembark.

```
QU YYZPWAC
.YYZOAAC 130953
FUEL ADJUSTED 9 PERCENT ABOVE AVG CONSUMPTION FIN 216

FP 0199/13 A320 216 C-FTJR SEL/FQCP C 30  CAPT GM MURPHY

RMKS
ADTNL FOR POSS ENRTE WXX/CBS

CYYZ/05 1100Z PEAR8 CALON V36 YVV J531 SSM J500 VLN J504 YEA J476
ALOMO DARKE7 CYYC/34

ALTN CYEG  G/C 1451
```

TO	EET	ET/AT	F/L	M/T	T/T	DST	TAS	TDV	WIND	COMP	TR	MFOB
YVV	0016.../..		CLB			94		P09	25052		42	11.0
T O C	0023.../..		CLB			47		P05	24060		41	10.6
APNEL	0025.../..		350			13	457	P05	24060	M029	41	10.5
SSM	0040.../..		350			108	455	P04	24048	M025	40	9.9
YQT	0113.../..		350			236	453	P02	28032	M023	39	8.4
VBI	0141.../..		350			197	455	P02	26049	M040	40	7.2
VLR	0208.../..		350			189	453	P03	21059	M029	40	6.0
VLN	0241.../..		350			236	451	P01	20054	M025	35	4.6
YEA	0308.../..		350			194	453	P04	31036	M015	33	3.5
FILLE	0320.../..		350			86	459	P08	32071	M048	35	3.0
T O D	0321.../..		350			1	457	P07	32071	M040	35	3.0
ALOMO	0324.../..		DSC			23		P04	31046		36	3.0
CYYC	0341.../..		DSC			73					38	2.6

```
                     TTL DST 1497 TMP P05   CMP M025

BURN  FIT  ALTN  CF   WXX  TF   ELW
10.0  1.2  1.4   0.3  0.6  0.3  58.1

ALTN CYEG
CYYC BARBOW7 CROSY V301 YEG DCT CYEG
```

TO	EET	ETA	F/L	DST	TAS	TDV	WIND	COMP	TR	BURN	MFOB
CYEG	0031	.....	160	146	342	P01	30015	M006	37	1.4	1.2

Page one of a Toronto-to-Calgary flight plan

All commercial aircraft must receive clearance before they can take off. Many airlines have their own flight dispatchers who, in conjunction with the crew, decide the airplane's flight plan (its routes and flying altitudes). The flight plan is then filed with air traffic control, which approves or modifies it and gives the crew final instructions prior to departure. Pilots follow these to the letter, checking clearance information against their charts and maps. (Now you know what they carry in their flight bags.)

Pilots are also responsible for inputting a four-digit transponder code into a device located in the aircraft's flight deck. The code allows the computers at air traffic control to interface with the airborne aircraft, thereby obtaining information about altitude, flight path, and speed, which is then displayed on the controller's radar screen.

The busiest airport in the world for movements (which are take-offs and landings), with over 980,000 a year, is Atlanta's Hartsfield-Jackson International, which handles 84 million passengers. Toronto's Pearson International Airport, the busiest in Canada, has 418,000 aircraft movements in a year. The expanding number of flights worldwide has prompted the development of many innovative strategies and technologies to keep order in the crowded skies.

For example, ever-increasing traffic has led to the reduction of lateral and vertical distances between flying aircraft; various procedures and technological advances have allowed this to occur without jeopardizing safety. One such device is the onboard Traffic Alert and Collision Avoidance System (TCAS), which communicates with other aircraft and determines if a nearby airplane is a threat.

So the next time you look out an airplane's window at the clouds below, remember you are on a highway like any other, but with one important difference: You are surrounded by a large and complex network of highly skilled people, all working to get you safely to your destination.

During the initial level-off, a few cruise checks are done and, more importantly, we divvy up the crew rests. Four pilots are required for this long-haul flight, one captain, two first officers, and a cruise pilot. The cruise pilot has calculated we have 14 hours before the top of the descent (when we start to move down), so we collectively decide the breaks will be split up three and a half hours each. The captain and I opt to fly shifts one and three, so the other two go back and either retire in the bunk beds directly behind the flight deck, or enjoy a relegated J-class (business) seat and perhaps watch a movie.

Most parameters, like pressurization, temperature, engines, electrical systems, and hydraulics, are monitored by the aircraft itself, freeing up the pilot from many obligatory duties — as a result, the odd newspaper may be read (but you didn't hear that from me, since reading for pleasure in the flight deck is forbidden). Another surprising aspect of the flight deck is controlled rest. Studies from NASA have proven brief naps in the flight deck are extremely beneficial, so the policy has been recently implemented. However, many restrictions and procedures still exist, so don't worry that both pilots are up there snoring away.

System after System

➤ Number of aircraft systems categorized in aircraft manuals: up to 24.

➤ Manual that pilots, maintenance crew, and flight dispatch consult to ensure all systems are a go: Minimum Equipment List (MEL).

➤ System used to move or operate several aircraft devices: hydraulics.

➤ Number of hydraulic pumps for A340: 8 (4 engine driven, 3 electric, and 1 ram air turbine).

➤ Main sources of electrical power: engine generators, Auxiliary Power Unit (APU, located in tail), external (plug-in), batteries.

➤ How landing gear works: electrically activated and hydraulically operated.

➤ Weight of A340 landing gear: 18.7 metric tons (20.6 tons).

➤ Where to view A340 landing gear: Science Museum in London, England.

➤ What aircraft tires are inflated with: nitrogen.

➤ Number of oxygen systems in aircraft: 3 (cockpit, cabin, and portable oxygen).

➤ Maximum water capacity for A340-500: 282 gal. (1067 L).

➤ Waste capacity for A340-500: 370 gal. (1400 L).

➤ Examples of exterior lights: landing, navigation, strobe, wing inspection, taxi, anti-collision.

➤ Typical height that landing lights are turned on before landing: 10,000 ft. (3048 m).

➤ Location of fuel: in wings and sometimes in center belly and tail.

➤ Fuel capacity for A340-500: 168,000 kg (370,377 lb.).

➤ Main navigation tools: laser gyros for inertial navigation (to detect tiny shifts in momentum) and GPS (Global Positioning System).

➤ Advantage of laser gyro over conventional spinning gyro: no moving parts.

➤ Accuracy of GPS: within feet.

When pilots have a complaint with the airplane they write it in the logbook as a snag. Here are some of the comical ones.

(Problem) Left inside main tire almost needs replacement.
(Solution) Almost replaced left inside main tire.

(P) Test flight OK, except auto-land very rough.
(S) Auto-land not installed on this aircraft.

(P) Something loose in cockpit.
(S) Something tightened in cockpit.

(P) Evidence of leak on right main landing gear.
(S) Evidence removed.

(P) Dead bugs on windshield.
(S) Live bugs on order.

(P) Aircraft handles funny.
(S) Aircraft warned to straighten up, fly right, and be serious.

(P) Radar hums.
(S) Reprogrammed radar with the words.

(P) Unfamiliar noise from engine.
(S) Ran engine continuously for 4 hours, noise now familiar.

Pilots discuss speed in units of miles per hour, kilometers per hour and, as you now know, knots. Another aviation unit we use is Mach speed, or the speed of sound.

The Sound of Speed

More than half a century has passed since Chuck Yeager broke the sound barrier on October 14, 1947. He was flying the Bell X-1, an airplane that looked like a bullet with wings. With all the advances in aviation science since then, it's a wonder that more airplanes aren't supersonic. But sound is more than a speed marker. It's also an important factor in how modern airplanes are made and flown.

The speed of sound changes at different altitudes, depending on variables such as temperature and air density, so there is no set speed in miles or kilometers per hour that corresponds to the speed of sound. This is why physicist Ernst Mach (pronounced "mock") devised a system to describe the relationship between an object's speed and the speed of sound. If an object is traveling at Mach 1 it's traveling at the speed of sound, Mach 2 means twice the speed of sound, and so on.

When an airplane moves at supersonic speeds, waves of air pressure pile up ahead of the airplane and compress, forming shock waves. These are similar to the bow waves created by a boat as it moves through water. The shock waves ripple out and away from the aircraft, toward the ground. When a shock wave hits your eardrum there is a sudden change in pressure called a sonic boom.

The retired Concorde traveled up to twice the speed of sound and was the only commercial passenger airplane that flew at supersonic speeds. Because of the sonic boom it created upon breaking the sound barrier, the Concorde was restricted to flying over the Atlantic Ocean between New York and Europe. Usually its sonic boom could only be heard in southwestern Nova Scotia.

Sound can have a major impact on people on the ground, which is why sound reduction is a major initiative in the aviation industry. Currently, Air Canada's fastest airplane is the Airbus A340, with an average cruising speed of Mach 0.82, but the B747 blows by us doing on average Mach 0.86. It doesn't seem like much, but for a 10-hour flight, it means about 24 minutes. (Our new kid on the block, the B777, hits Mach 0.84.)

These days, most airports also have strict noise-abatement procedures, which are outlined on pilots' charts. When flying out of Toronto Pearson International Airport, for example, pilots can't turn until they're 3000 feet (915 m) above the ground. Takeoffs and landings are also forbidden between 12:30 a.m. and 6:30 a.m. Some

The retired Concorde at Heathrow

of Air Canada's flights, however, have special exemptions because of the quiet airplanes used.

"Flight 2341, for noise abatement, turn right 40 degrees."
"But Center, we are at 35,000 feet. How much noise can we make up here?"
"Sir, have you ever heard the noise a B747 makes when it hits a B737?"

For existing aircraft, Canadian and American regulators have been implementing graduated noise-level reductions for jet engines. We are now into stage III, which requires airlines to ground many old airplanes or retrofit them with "hush kits" to control noise.

In today's quiet flight decks, a headset is optional as the only sound heard is the slipstream of air against the windscreen. Cabins, too, have become increasingly quiet, which is a big hit with passengers.

✈ ✈ ✈

Since I've been with Air Canada, I have seen several airplane types fly their last commercial flight. The fleet is always changing to make room for the new arrivals. Each month, Air Canada sees new airplanes land from Brazil — the Embraers. A new feature popular with passengers is the touch-screen, on-demand video players at each seat.

A new heavy hitter, which will have touched down by the time this book is on bookstore shelves, is Boeing's 777, the Triple Seven. The long-range version of the B777 is capable of flying non-stop between any two points in the world. In 2010, Air Canada's fleet will be joined by the new, state-of-the-art way of thinking — the Boeing 787, a.k.a. the Dreamliner. The Dreamliner was designed by engineers thinking "outside the box." One neat feature is the addition of a window to the lavatories. What about when the plane is on the tarmac and the ground crew is lurking about? Well, the clever engineers put sensors in the landing gear that send a signal to black out the windows in each lavatory — since all the windows in the Dreamliner are electronically shaded.

So what do airlines do with outdated planes? Unlike cars, which are thrown on the scrap heap, retired planes are treated with a more sensitive approach. Just a couple of years ago, the family and I were in Arizona and I wanted to visit one of the largest graveyards for airplanes. We were denied entry. I guess they thought a family of five wasn't a serious buyer of an old B747. (Airline passes are yet another fringe benefit of the job. They do, however, come with stipulations, and one of them is flying on a stand-by basis. We certainly felt the ramifications of that when trying to get back to Toronto after a major Canadian airline went belly up.)

Airplane Heaven

Glinting under the hot sun in the Arizona desert is FIN 714, an Air Canada DC-9 known simply by its flight identification number. After 33 years of service, it now sits parked in a compound just off Interstate 10, northwest of Tucson, among hundreds of other retired DC-9s, jumbo jets, and aircraft.

The Evergreen Air Center is the largest commercial aircraft storage base in the world, capable of holding up to 300 aircraft. Air Canada has several of its DC-9s and Boeing 747s at this peaceful facility, where they rest alongside aircraft like L-1011s and DC-8s. Unlike Canada with its harsh climate, Arizona and other locations in the southwestern United States are ideal for storing aircraft outdoors, thanks to the consistently warm, sunny weather and dry, light winds.

The airplanes put into long-term storage here aren't just left to bake in the sun among the rattlesnakes and tumbleweeds. First, they are embalmed. Qualified aircraft-maintenance personnel empty fuel from the tanks, remove oil from the engines, and drain hydraulic fluid that once operated the landing gear and control surfaces. Instrument probes are also plugged, and expensive instruments are covered or removed. All the while, the records for each aircraft are carefully filed and stored.

Airplanes are put out to pasture for various reasons. Sometimes, an economic downturn means less air traffic, so airlines will put some airplanes into temporary storage until the economy picks up. Old aircraft still in perfectly good shape may also be retired to make way for newer airplanes that are larger, faster, and more fuel efficient. The Airbus A340, for example, is not only less noisy and more comfortable for passengers than the Boeing 747 it replaced, but it is also 60 percent less expensive to operate.

Many airplanes that end up at the Evergreen Air Center never leave. Some are stripped of their parts, which help keep other airplanes flying. Others are dormant for years and then eventually fall victim to the welder's torch. But the lucky ones are sold to other airlines. In fact, many of Air Canada's well-maintained aircraft return to service, operating for other carriers all over the globe. Airplanes expected to reside at the storage facility for only a short stint undergo regular engine starts and various electrical and system checks. That way, prospective buyers can have their airplane up and running in days. Some other lucky planes are preserved forever by playing key roles in films and on television shows. For ABC's hit drama *Lost*, a retired airplane was pulled apart and set down on a Hawaiian beach to stand in as the ill-fated Oceanic Flight 815 aircraft.

Today Air Canada's FIN 714 sits in limbo, awaiting its fate. During its tour of duty it performed 67,657 landings and amassed an incredible 75,826 flight hours. A pilot who is also retiring with more than 30 years of service might accumulate a mere 20,000 flight hours. Still, there is always a possibility that FIN 714 will fly again one day, and that its new owner will be just as proud as its original owner was more than three decades ago, when the shiny new DC-9 first took to the air.

Through data link we get the latest weather for our route and Hong Kong itself. During flight planning, we noticed one of the coded aviation forecasts mentioned thunderstorms would be an issue in Hong Kong for most of the day. Our arrival time, still some 14 hours away, will have us landing when thunderstorms are at their peak, in the afternoon. Thunderstorms are avoided the same way mariners avoid icebergs and motorists avoid 10-foot-deep potholes.

Stormy Weather

With jagged zigzags lighting up the sky and tremendous, rumbling crashes, thunderstorms provide one of the greatest shows on earth. But the last thing pilots want is to get too close, so we avoid flying through them at all costs. When thunderstorms are prevalent, we keep an especially close eye on developing storm systems, as on a recent flight I piloted from Toronto to Calgary.

While perusing the flight plan that morning, the captain and I saw a reference to thunderstorms. We called the dispatcher handling Prairie-bound flights for a briefing. The weather office had forecasted thunderstorms developing in southern Saskatchewan that afternoon, so we boarded extra fuel in case we needed to change course. By noon, the thunderstorms had started to show up on the Canadian weather radar network. Typically, only 10 to 12 miles (16–20 km) wide, thunderstorms can form in clusters or lines that extend for hundreds of miles, and their heights can far exceed the altitudes an airliner flies. To steer us clear of these roadblocks, our flight path routes us along the northern United States.

A thunderstorm cloud is called a cumulonimbus, or Cb. Thousands of Cb clouds form every day around the world, with 2000 booming at any given time. They come with heavy showers, lightning, turbulence (see page 82), and are even capable of covering an aircraft in ice. Thunderstorms pack an enormous amount of energy. With more than 100 million volts of electricity, lightning strikes travel either between clouds or from a cloud to the ground, and vice versa. The air near a lightning strike is heated to 28,000°C (50,432°F), several times hotter than the surface of the sun! The rapid heating and cooling of air near the lightning channel causes a shock wave, which results in thunder. Texas, Oklahoma, and Kansas hold the title for most thunderstorms in the United States. In Canada, Ontario lays claim to the most thunderstorms annually.

Three ingredients must be present to breed a thunderstorm: moisture, a lifting agent, and unstable air. On the day of our recent flight, the lifting agent was a cold front sweeping through the prairies coupled with daytime heating, which caused moist air near the surface to rise quickly. Through its upper-air weather balloons and computer-generated weather charts, Environment Canada knew that all the elements were in place for thunderstorms. Overhead satellites and state-of-the-art radar, along with ground weather observations and pilot reports, confirmed their suspicions.

Once airborne, we crossed over into American airspace. The flight was smooth and our navigation computers indicated that we would arrive on time. As we skimmed the American–Canadian border, thunderheads started appearing on the horizon some 200 miles (322 km) to the northwest. We flicked on our aircraft weather radar and started "picking them up" (aviation lingo for detecting storms on radar). Our cruising altitude was 36,000 feet (10,973 m) above sea level, and we estimated the top of the thunderheads to be some 10,000 feet (3048 m) above us. By 4:00 p.m., the peak time for thunderstorms, we could see a billowy expansion, indicating very strong updrafts. An anvil-shaped cloud formation had developed on top, indicating that the storm clouds have poked into the next layer of the atmosphere, called the stratosphere (see also page 93).

Even 25 miles (40 km) away (the recommended safe distance), these thunderstorms were truly awe inspiring, so we made an announcement for passengers on the right side to have a look. Air traffic control had assured us that flight conditions would be smooth on the route we took and, sure enough, the seat belt sign stayed off the entire time. The light show was an added treat.

But what happens if lightning does hit the airplane? Many passengers envision a wing snapping off during a lightning strike as depicted by Hollywood B-movies, but it's not so. Usually lightning enters at one part of the plane and exits at another with no damage. Sometimes it does leave little pin holes and this is why maintenance gives a thorough inspection after a lightning strike.

Red Alert!

During your travels you may hear the airport is under red alert. This indicates thunderstorms are in the immediate area. A device situated on the airport premises detects thunderstorms close enough to do harm, meaning operations on the ground must come to a halt. Ground handlers, caterers, and fuelers all must take cover and wait out the storm to avoid lightning strikes. It leaves some passengers scratching their heads, especially if no showers are present. Just recently, after an eight-hour flight from London, England, to Toronto, we were short of the gate by 100 feet (30 m) when a red alert was announced. We had to wait because no one could marshal us in, work the Jetway, or unload the bags, which translated into a wait of 30 minutes. . . . So close and yet so far.

Going for a Ride . . . In a Thunderstorm?

It's engrained into every pilot: Thunderstorms and airplanes don't mix! It starts in initial flight school and is reinforced throughout a pilot's career. Taking on a thunderstorm means a pilot will invariably go on the ride of his or her life.

Some Flights That Didn't Make it

Even though much is known about thunderstorms, they still pluck airplanes out of the sky. In 1985, the engines on Delta Airline's L1011 couldn't spool up fast enough to recover from severe low-level wind shear (LLWS) on approach into Dallas, Texas. This tragic lesson brought to the forefront the phenomenon of LLWS (see page 83). Following this accident, many large airports in the U.S. and around the world are now equipped with an LLWS-detection system that gives pilots an advance warning. Actual data taken from the accident was implemented in simulators for pilots to hone their ability to escape LLWS; however, only those who knew it was coming and flew the aircraft to its limits made it. Fortunately, new airliners are built with onboard LLWS-detection systems. The jumbo jet I fly starts yelling "wind shear, wind shear," providing a life-saving heads-up.

In Little Rock, Arkansas, in 1999, American Airlines MD80 skidded off the end of the runway while landing in a well-developed storm. This crash was another grim reminder that thunderstorms can deal out fatal rides.

Over the Top
In an attempt to dodge thunderstorms, some pilots elect to fly over them when they are thought to be a safe distance below, yet thunderstorms can rapidly raise their ugly heads and give an aircraft an unforgettable ride. Many airliners are restricted to a maximum altitude of 40,000 feet (12,192 m), so flying over the big boomers is not an option. The spread between VMO (the maximum allowable speed for a particular airplane) and stall speed (the speed at which lift ceases) can quickly close in on each other. It's not a nice feeling to be high up in the thin air, where flight controls are less effective and where turbulence can push an airplane into forbidden flight envelopes. No wonder this narrow airspeed gap is referred to as the "coffin corner" or "coffin alley."

What Goes Up Must Come Down
Not only have pilots gone for extraordinary rides, but it's been known to rain tadpoles, fish, and even skydivers after they've been sucked into updrafts — just like the storm that ingested Dorothy and Toto in *The Wizard of Oz*. In June 2002, skydiver Mathieu Gagnon was pulled into dark storm clouds while parachuting in Ontario, Canada, and was sucked up by the storm. After rising almost 3300 feet (1006 m), he cut away his main parachute and fell out of the ominous clouds. Using his reserve chute, he safely touched down about 15 miles (24 km) south of the airfield where he was supposed to land.

In the Name of Science
John Aiken, a research pilot at the NRC (Canada's National Research Council) in Ottawa says, "Research flights did exist back in the '70s and '80s that purposely ventured in Cbs [thunderstorm clouds], but nowadays we try to avoid them at all costs." Recently, in collaboration with Environment Canada, John flew the NRC's Twin Otter to collect data from lake breezes off the Great Lakes, which are known for thunderstorm development. Environment Canada's computerized weather models still thirst for data to get a handle on thunderstorms, but John ensured the research flights were at a safe distance.

John's coworker and another research pilot, Tim Leslie, had this to say about thunderstorms: "In the mid-80s I was flying a T-33 from Cold Lake, Alberta, to Moose Jaw, Saskatchewan, and inadvertently entered a thunderstorm. It was a helluva ride! The VSI [vertical speed indicator] was pegged in both directions and the aircraft was kicked around like a leaf. Hail ripped off all the navigation lights and did a pretty good number on the leading edge. However, throughout the entire ordeal, the ever-dependable Rolls-Royce engine just chugged away, ingesting and spitting out gallons of water. Regarding rides, the trip through the thunderstorm was far more harrowing than any hurricane trip I have done."

On the Lookout North of the Border

Keeping an eye on thunderstorms in Canada — from a safe distance — is Environment Canada, with its pumped-up Doppler radar system. Steve Lapczak, responsible for the 30 radar site upgrades says, "They have all been converted . . . ahead of schedule and on budget." Doppler radar was implemented in the U.S. during the 1990s.

What can Doppler radar do that the older system couldn't? Until recently, weather radar could only detect the size, shape, intensity, and direction of movement of thunderstorms by tracking the precipitation within it. Now with the advent of Doppler radar, rotating motions inside a convective cloud can also be determined, possibly foreshadowing the onset of a tornado. Doppler allows meteorologists to investigate the wind patterns within storms, which detects downbursts, low-level wind shear (LLWS), and microbursts more accurately (for more on LLWS see page 83). One lead scientist in the group says, "Doppler compared to the old system is like CDS compared to vinyl discs." The next-generation radars are able to determine the type of precipitation (snow from rain from hail) using a state-of-the-art system referred to as "polarization diversity."

With 2000 thunderstorms booming around the globe at any given time, it's inevitable a pilot and a Cb (cumulonimbus cloud) will cross paths. Sometimes these paths come close when aircraft are being vectored around convective weather by air traffic control (ATC). In the past, ATC had dubious information on the exact whereabouts and details of thunder-bumpers, and they relied on the information given to them by what the pilot "painted" on the onboard weather radar. (Incidentally, many small aircraft do not have weather radar capability.) Every so often vectors and holding patterns got a little too close for comfort, but Lapczak says that in the very near future ATC will have technology that will allow them to superimpose detailed Doppler information onto their screens, ensuring the pilot's ride is a smooth one.

Going for a ride in an airplane should be enjoyable. It shouldn't include horrific stories involving "rides of your life" in or alongside perilous thunderstorms.

➤ Number of tornadoes in the U.S. per year: 800.
➤ Canada ranks second in the world with 400 per year.
➤ Kansas, Texas, and Oklahoma hold the title for most thunderstorms in the U.S. per year. For Canada, Ontario takes the prize for most thunderstorms.

The captain I am flying with is easy to talk to — something you appreciate during long-haul flights. Think about it, you meet a person for the first time, knowing you are going to spend hours sitting next to each other in a room not much bigger than a closet. Think for a moment about the people you know. Who would you easily get along with in a flight deck environment? Who would be some of the rather-nots?

The conversation begins on the topic of where we live. Because postal codes vary so much among pilots, it's always interesting to discover where everyone lays their hat. I mention I'm living in Halifax, and immediately he asks about the impact Hurricane Juan had in 2003. I told him I live one minute from Halifax's Point Pleasant Park, where you can still see the devastation and scars from when Juan visited Halifax.

Weather is a major consideration for any airline. Here in North America we get it all, from freezing rain in the winter to humongous summer thunderstorms. We have to contend with fog on the east coast, mountain waves over the Rockies in the west, snowstorms, high winds — you name it. Because Air Canada is an international airline, it's inevitable we pilots will experience most of what Mother Nature can dish out. One big weather maker that always makes the news and plays havoc with schedules is hurricanes. I worked at Canada's Hurricane Center in Halifax years ago, and these nasty storms certainly kept people on their toes.

'Tis the Season for Hurricanes

The peak month for these self-sustaining heat engines is September, closely followed by August and October. Tellingly, its name is derived from "Hurrican," the Carib god of evil. This cyclonic, twisting Goliath requires a combination of conditions to rear its ugly head.

First off, sea surface temperatures must be at least 27°C (81°F). Hurricanes tend to form five to eight degrees north or south of the equator, since the rotation begins by the influence of the Coriolis force. (Remember that phenomenon that causes everything in the northern hemisphere to be deflected to the right but has zero effect at the equator? See page 89). A weak weather disturbance — a tropical wave moving westward from the area of Cape Verde, Africa — is then needed to mix all these ingredients together. This atmospheric support acts as an impetus for thunderstorm formation, grouping this newborn storm into a self-perpetuating heat engine, as latent (hidden) heat is released from the process of condensation, causing more vertical development.

Hurricanes are born as a tropical depression (or a "baby" low pressure system) with winds less than 34 knots (39 mph or 63 km/h). The next stage, the tropical storm, has sustained winds of 35 to 64 knots (40–74 mph or 65–119 km/h), and marks when the storm is christened with an official name. By definition, a tropical storm becomes a hurricane when its winds exceed 64 knots (74 mph or 119 km/h). Today, names alternate between the male and female gender and are predetermined for six years, after which time the names are repeated. The list contains a name for each letter of the alphabet. (Q, U, X, Y, Z are not included because of the lack of names for those letters.) Some names are retired because of their destructive past. Andrew (1992), Juan (2003), and Katrina (2005) were pulled from the roster.

The most significant clouds are the cumulonimbus (known as thunderstorm clouds), which top out at a whopping 50,000 to 60,000 feet (15,240–18,290 m) above the ground. Hurricane winds as high as 130 to 150 knots (150–173 mph or 241–278 km/h) have been reported — any higher and wind-measuring devices (called anemometers) are destroyed. However, engineers estimate that winds of 170 to 200 knots have occurred, based on their observations of structural damage. Winds quickly drop off to less than 15 knots (17 mph or 28 km/h) in the eye of the storm. The diameter of the eye varies from approximately 10 to 40 miles (17–64 km), and in this region there is usually no precipitation and very few clouds. The eye is the result of subsidence: the air at the center of the storm sinks, which creates compression, thereby warming the air and breaking up clouds. Temperatures at the core tend to be five or more degrees Celsius warmer as a result.

Just to give some indication as to the intensity of these storms, over a period of 24 hours a small hurricane will drop 20 billion tons of water. They possess the equivalent in energy as 500,000 Hiroshima-sized atomic bombs, averaging six atomic bombs per second. Typical wave heights get up to 35 to 40 feet (11–13 m), although the very intense hurricanes generate waves of 45 to 50 feet (14–16 m), with some reaching heights of over 98 feet (30 m).

Hurricanes in the tropics typically move westward at speeds of eight to 16 knots and then curve northward as they are steered by the clockwise winds around the Bermuda High. As they track northward they will usually show signs of fading as they encounter colder waters. Furthermore, when they track over land the surface friction tends to take the wind out of their sails.

There are five categories with five being the most intense, packing winds of 135 knots or greater.

The Saffir-Simpson Hurricane Scale

Category One (64–82 knots, 74–95 mph, or 119–153 km/h): No real damage to building structures.

Category Two (83–95 kt, 96–110 mph, or 154–177 km/h): Some roofing material, door, and window damage to buildings. Considerable damage to shrubbery and trees. Some trees blown down.

Category Three (96–113 kt, 111–130 mph, or 178–209 km/h): Large trees blown down. Mobile homes and poorly constructed signs destroyed.

Category Four (114–135 kt, 131–155 mph, or 210–249 km/h): Shrubs, trees, and all signs are blown down. Complete destruction of mobile homes. Extensive damage to doors and windows.

Category Five (135 kt, 156 mph, or 249 km/h and greater): Complete roof failure on many residences and industrial buildings. Some complete building failures with small utility buildings blown over or away. All shrubs, trees, and signs blown down. Complete destruction of mobile homes. Hurricane Katrina (August 2005) was a category five when it hit southern Louisiana, Mississippi, and Alabama.

Now and again Mother Nature throws a curveball, and this is what happened in late October 1991. Hurricane Grace moved northward on October 27 and became extratropical. (Extratropical means beyond or outside the area of the tropics. Not to be confused with extra, which means over and above what is normal.) It was a combination of extenuating circumstances, and it caused the rebirth of a massive destructive monster, which the National Weather Service described as the perfect storm. A weak low-pressure system sitting off Nova Scotia combined with a cold intrusion of air from the continent and the moisture from Grace, which resulted in a storm that caused millions of dollars in damage along the eastern seaboard. Also dubbed the Halloween Storm, it sank a sword-fishing vessel, the Andrea Gail, with all hands on board perishing. It became the basis for Sebastian Junger's novel, *The Perfect Storm*, which was made into a movie. Mother Nature threw yet another curveball when the storm found a warm-water source from the Gulf Stream, which sent it back into hurricane status. The Halloween Storm, or Perfect Storm, was then known as the "Unnamed Hurricane." A decision was made not to give it a name, because the media's attention was still focused on the enormity of the damage, and it was felt renaming would cause confusion. The Unnamed Hurricane brought a bizarre ending to one of the most massive Atlantic storms on record.

This redevelopment of hurricanes also happened in October 1954. Hurricane Hazel first struck the Carolinas as a category four. She merged with a low-pressure center to become a colossal extratropical storm that devastated Toronto, causing 76 deaths.

We have come a long way in forecasting the movement of hurricanes and in understanding the complexity of these one-eyed monsters. And yet, only Mother Nature herself knows where the curveballs she throws next will land.

Thankfully due to their slow movement in the tropics and the advanced detection satellite and aircraft reconnaissance provide, hurricanes are not a surprise to weathermen, pilots, or flight operations. But because of the size of these storms (with widths measured in

Hurricane Katrina damage in the hard-hit Ninth Ward, New Orleans

hundreds of miles), they tend to wreak havoc for long periods of time and affect a large area — for the airline industry, this means airport closures or major flight delays. A pilot must carry extra contingency fuel in case he or she must fly to an airport a safe distance away from the storm. Power outages also accompany hurricanes so airports in an affected area may operate on standby power or with a skeleton crew — conditions conducive to delays and cancellations. Hurricane Katrina knocked out oil rigs in the Gulf of Mexico causing oil prices to rocket sky-high, affecting airlines' bottom line. Clean up after a hurricane is also an issue, making getting to and leaving the airport a concern for passengers.

Tired of your job? What about working on weather reconnaissance flights that fly into the eyes of hurricanes in the name of science and research? Known as Hurricane Hunters, pilots from the NOAA (National Oceanic and Atmospheric Association) and American Air Force Reserve fly two P-3 Orions and C-130 Hercules respectively, all laden with meteorological instruments to compile data from specific storms for research.

Many confuse the term tornado with hurricane. Tornadoes are a direct result of a thunderstorm with dimensions hundreds of feet wide; they last only minutes and form over land. A hurricane, on the other hand, has a berth of hundreds of miles, travels large distances, lasts for days, and forms over warm ocean water. One reason for the confusion is that the word cyclone was used in the United States for a tornado, and a hurricane that forms in the Indian Ocean or certain parts of the Pacific Ocean is also called a cyclone. The true definition of cyclone is anything that spins counterclockwise in the northern hemisphere. A hurricane is called different things depending on its birthplace. In the Atlantic, it's called a hurricane; in the western Pacific, typhoon; in the Indian Ocean and eastern Pacific, tropical cyclone or cyclone; and the Aussies sometimes call it "wily-wily."

The flight today takes us over northern Ontario, Hudson Bay, Baffin Island, the west coast of Greenland, adjacent to Canada's most northern settlement, Alert, and about 60 miles (97 km) from the North Pole, where we pick up one of the four polar routes over Russia. We will pass through Russia's, Mongolia's, and China's airspace, and then into Hong Kong. Pilots reference feet for altitude, but as a rule, when flying in communist countries we fly referencing the meter. Luckily we have on board a button that easily does the conversion.

Mixed Measurements

At every airport in the world, from Rome to Rio de Janeiro, pilots and air traffic controllers communicate in English. But when it comes to the unit systems they use for crucial information like altitudes, airspeeds, and distances, it's not that simple. Pilots constantly shift from one system to another, using a mix of metric-derived si units, imperial measurements, and naval standards that to a layperson may seem confusing. Fortunately, in today's modern cockpits conversions are easily made at the push of a button, which means that the books of conversion tables we carry on every flight are rarely needed.

On a recent flight, while traveling through Russian airspace en route from Toronto to Tokyo, I had to convert the airplane's altimeter reading from feet to meters. Formerly communist Russia and communist China are unlike the rest of the world in that they use the si system for altitude. Altitudes are expressed primarily in feet — even in France, where the metric system was created, and where it and has been in use for more than 200 years.

Worldwide, countries are slowly moving toward adopting the metric system, but each country is changing at its own pace, so pilots flying internationally need to be aware of local norms. Metric is now standard for many measurements, such as visibility on the ground, which outside of North America is measured in kilometers and meters. Canada and the United States, however, often use feet or the statute mile (which equates to 5280 ft. and 1.6 km) for ground visibility, and the nautical mile (which equates to 1.15 regular miles and 1.9 km) is usually reserved for measuring distances. To further complicate matters, the speed of an aircraft or of the wind is often still measured in knots, originally a naval term meaning nautical miles per hour. (Doubling the speed in knots gives you the approximate speed in km/h, so 500 knots is about 1,000 km/h.)

When it comes to fueling up, both gallons and liters are used, and computers convert the fuel volume into weight. The weight is measured in either kilograms or pounds, depending on the aircraft type.

Canadians have been receiving weather reports in Celsius for decades, but Americans are just starting to switch over. U.S. weather reports have recently begun to use Celsius, but these values are usually still converted to Fahrenheit for the public. Pilots also need to keep track of two different outside-temperature measures in the flight deck. One is the true ambient temperature outside, and the other is a value that is warmer due to the friction caused by the airplane cruising along at nearly 10 times the average speed of a car. This temperature is called total air temperature, or TAT.

Units of air pressure are usually expressed in metric kilopascals, but pilots generally speak of inches of mercury or millibars. Air-pressure values are crucial in regular flight deck operations, as altimeters must be adjusted to local atmospheric pressure to determine the airplane's exact height above sea level.

According to the Montreal-based International Civil Aviation Organization, non-si units, like nautical miles, knots, and feet, are still permitted in aviation because they are so commonly used that changing them would require a major international effort. The goal, however, is to eventually create a more uniform international standard. Who knows? If most of Europe can agree to adopt a common currency, perhaps one day everyone will also agree on how to measure things. But for now, switching between unit systems is an important part of the airline pilot's job.

Conversation overheard between a London terminal controller and an American pilot:

ATC: N12345, descend to 3000 feet on 1019 millibars.

N12345: Could you give that to me in inches?

ATC: N12345, descend to 36,000 inches on 1019 millibars.

Pilot Math 101

Can't get your head around all these units? Here's some math to ponder.

➤ Formula to determine distance in nautical miles seen from the airplane: take the square root of altitude in feet and multiply by 1.23.

➤ How far can you see at an altitude of 37,000 feet? 237 nautical miles (approximate distance between Ottawa and Toronto).

➤ Rule of thumb to decide when to descend: multiply first 2 numbers of altitude by 3.

➤ How many nautical miles from your destination do you need to start descent at an altitude of 30,000 feet? 90.

➤ Meaning of the basic "one-in-60" navigation rule: if you are one degree off course, you will be one mile off course after flying 60 nautical miles.

➤ How far off course are you if you are 10 degrees off course after flying 30 nautical miles? 5 miles.

➤ Formula for converting volume of fuel (in liters) into weight (in kg): multiply liters of fuel by 0.8.

➤ How much does 100,000 liters of fuel weigh? 80,000 kg.

➤ Engine oil requirement for an Airbus A340-300: minimum of 10 quarts plus 0.6 quarts for each hour.

➤ How many quarts of oil would an A340 need for a 10-hour flight? 16.

➤ Time to climb 10,000 feet at 2000 feet per minute: 5 minutes.

➤ If a pilot descends from 10,000 feet at 1000 feet per minute, how long will it take to reach sea level? 10 minutes.

Let's take a break from all those calculations and instead take a peek at what's outside our window.

Night Flight

During a daytime flight, your airplane window affords many interesting bird's-eye views, but if you know where to look, night flights
yield even more spectacular sights. Along with the stars and con-
stellations, the dark offers its own amazing phenomena.

The planet Venus is the brightest object in the sky, aside from
the moon. Called both the morning star and evening star, it can
take on many guises. Atmospheric effects from Earth cause it to
flicker, changing its color from green to red to orange to blue. It
can look like a bright light that is following you. It's not surprising,
then, that it's been mistaken for a UFO. Recently, while flying east
over Russia during the wee hours of the night, we spotted lights
ahead fluctuating between green and red. At first we thought it was
an airplane, but it began to climb higher above the horizon, confirming that it was in fact Venus staging a dazzling light show.

Now and again we are lucky and catch sight of a shooting star
illuminating the night sky. Once while we were flying from Toronto
to Montreal, a very bright light streaked past the airplane's windshield. Reports the next day verified that a meteorite had landed in
New York State, only a few hundred miles away.

If our routing takes us north, we sometimes see another of
Mother Nature's spectacular shows, the aurora borealis (or northern lights). When these dancing lights are at their most breath-
taking, we let our passengers know — if it's not too late at night that
is. Another phenomenon worth seeing is daylight giving way to
night. This sharp black edge, called the terminator, can be seen
during eastbound flights.

The moon can also put on great shows, changing size and color
depending on the clouds and atmospheric conditions. A halo effect
around the moon, known as holding water or grease around the
moon, is caused by high clouds laden with ice crystals, and it's a pre-
cursor of advancing weather.

On a clear night we can enjoy great sights on the ground as well.
When flying at cruise level over the east coast, for example, we can
see the lights of all the Maritime cities at once. But my most
favorite city for landing at night has to be Las Vegas. The multicol-

Day gives way to night: the terminator and the moon en route to Frankfurt

ored neon lights of the hotels make me feel like I'm landing at an amusement park. A close second is Tokyo. As we take off over the water, we can see literally hundreds of fishing boats, their lights blazing to lure the fish to the surface. Flying over Holland provides a view of hundreds of greenhouses, and part of the border shared by Pakistan and India is illuminated at night — an 1800-mile-long (2897 km) sight to be seen. There is no limit to the fascinating sights you can see from the air: dimly lit Arctic settlements during the long Arctic night, moonlight dancing across the snow cover or over lakes in the summer, fireworks from above, oil rigs burning off gas, the rising moon. . . .

A very welcome nighttime sight is the array of landing lights at the destination airport. The runway's brilliant approach lights are impressive, their different colors communicating information from the ground to the pilots (see page 63). When I was a flight instructor teaching landings at night, I asked the control tower to turn up the lights to full brightness, strength five. It was almost blinding from above. The controller jokingly made a comment about the power bill going up.

Colored lights are important markers on the body of an airplane as well. The international standard is green light on the right wingtip, red on the left, and white on the tail. Airliners also have white strobe lights to make them easier to spot. However, when we're in a dense cloud layer the strobes may be distracting to other pilots and passengers, so if the cloud layer persists, we might turn off the strobes until we're out of it.

Another advantage to night flight is the smoothness of the air. Much of the turbulence encountered at low levels is due to the heating of the sun. At night this process abates, giving way to smooth flying. For years I've flown under an amazing blanket of stars and I still marvel at what I see. So next time the cabin lights are dim, be sure to look outside.

Night Moves

➤ Number of degrees below the horizon the center of the sun must be for night to occur: 6.

➤ Name for the dividing line between day and night on a celestial body: terminator.

➤ Objects that pilots flying at night were once required to carry: a reliable timepiece and a functioning flashlight.

➤ Device that helps a pilot to locate an airport amid the ground lights of a community: airport beacon.

➤ System that allows pilots to activate airport and runway lights (found at smaller airports only): Aircraft Radio Control of Aerodrome Lighting.

➤ Typical height at which airliner's landing lights come on: below 10,000 ft. (3048 m).

➤ Aircraft that are prohibited from landing or taking off between 12:30 a.m. and 6:30 a.m. at Toronto Pearson International Airport: Airbus A320, A330, and A340, and Boeing 737, 747, and 767.

➤ Example of flights exempt from the above rule: AC156 (from Vancouver) and AC146 (from Calgary).

➤ Official authorized to give special permission to land or take off during curfew hours: the airport manager.

➤ Why cabin lights are dimmed for landing: to enhance night vision for passengers.

Two Christmases ago, like many before, I had the honor of working and flying over the North Pole on the way to Hong Kong. There is a special sight to keep watch for . . .

Santa the Aviator

Unlimited visibility prevailed in winter's starlit darkness on a Christmas flight over the top, non-stop from Toronto to Hong Kong. It offered a rare glimpse of the North Pole and I scanned the frozen terrain for hints to Santa's whereabouts. Though my last pilot medical still deemed my eyesight faultless, I can't be certain I did indeed spot the glowing lights of Santa's busy toy factory with elves working frantically inside. (On returning home, my six-year-old son was ecstatic to hear of my bird's-eye view.)

During my search, I pondered the similarities and differences between an airliner and Santa the aviator. Santa Claus prefers a snow-covered landing pad, but it's not so for airplanes. Pilots prefer runways to be bare, and if dubious conditions exist, ground equipment will leap into action, incorporating snow ploughs, sweepers, and vehicles to spread a noncorrosive chemical to melt snow and ice. We also consult charts to determine if the reported braking action and crosswinds are within limits; if not, it's off to another runway or airport for landing.

It's well known Santa lands on rooftops with very steep pitches. Our tolerance for takeoff and landing is a slope of two degrees.

Very sophisticated landing instruments, both on board and at the airport, are a must in order for pilots to find the runway in snowy conditions. I'm certain Santa doesn't navigate using global positioning satellites or laser-mounted gyros to detect momentum shifts. Does he navigate by dead reckoning, his magnetic compass rendered ineffective in extreme northern latitudes?

While traversing the North Pole our airspeed is about 83 percent the speed of sound. One website with statistics on Santa Claus postulated Santa's required speed must be 3000 times the speed of sound in order to reach everyone on Christmas Eve. He must also have special permission to bust the mandatory airspeed restriction of 250 knots (288 mph or 460 km/h) below 10,000 feet (3048 m) above sea level. Rules must also be twisted for Santa not having the appropriate navigation lights, although Rudolph's red nose could improvise for a red anti-collision light. There are also the deicing systems that are required as he enters cloud, and the gamut of instruments necessary to keep the sled upright in disorienting clouds. (A non-trained person flying in cloud is statistically proven to last under a minute before plunging into a spiral dive.) Department of Transport could easily ground Santa's sled on hundreds of violations, but in the name of Christmas spirit Santa has been given a special flight permit.

As you now know, in order to stay current, airline pilots must undergo rigorous testing every six months in a flight simulator (see page 8). For the pilot, one of the toughest hurdles to overcome during a check ride (a flight test) is loss of an engine on takeoff. It reminds me of a story that circulates in the aviation world about Santa's check ride.

Apparently even Santa Claus could not escape the required flight test with a Transport Canada flight inspector. In preparation, Santa had the elves wash the sled and bathe all the reindeer. Santa got his logbook out and made sure all his paperwork was in order. He knew the inspector would examine all his equipment and truly put Santa's flying skills to the test. The examiner walked slowly around the sled. He checked the reindeer harnesses, the landing gear, and even Rudolph's nose. He painstakingly reviewed Santa's weight and balance calculations for the sled's enormous payload. Finally, they were ready for the check ride. Santa got in and fastened his seat belt and shoulder harness and checked the compass. Then the examiner hopped in carrying, to Santa's surprise, a shotgun. "What's that for?" asked Santa in disbelief. The examiner winked and said, "I'm not supposed to tell you this ahead of time," then he leaned over to whisper in Santa's ear, "but you're gonna lose an engine on takeoff."

Recently, airliners around the world were equipped with devices to interrogate other aircraft in close proximity, determining direction and altitude; if need be, a resolution advisory is given (e.g., "climb," "descend," or "monitor vertical speed"). It's not known whether Santa's sled has been updated with this new technology, so it will be hard to see him coming. However, having flown many Christmas Eves, I can confirm that pilots generally make an announcement to advise passengers when air traffic control detects an unknown blip on their radar originating from the North Pole.

This Christmas Eve no doubt I'll be flying. (Generally speaking, junior crews both in the flight deck and cabin will have the distinction of working through Christmas.) If our flight plan takes us over any great vantage points to spot Santa and his hard-working reindeer, like Baffin Island or Greenland, the passengers can expect a briefing on Santa's whereabouts and a "Season's greetings" from the flight deck.

A Bird's-Eye View

Not many people have an office like a pilot's — one that travels at 500 knots (575 mph or 926 km/h) at 40,000 feet (12,192 m) with an outside temperature of –60°C (–76°F) and winds as high as 200 knots

View from above: a volcanic island north of Iceland (L) and the Rockies (R)

(230 mph or 370 km/h). While Air Canada flights don't get quite high enough for the pilots to see the curvature of the Earth, it was possible to view this spectacular sight from the flight deck of the supersonic Concorde as it soared along at 60,000 feet (18,288 m).

As a point of interest, "flight deck" is the modern term for cockpit, which was adopted from an old nautical term used to describe a space toward the stern that's lower than the rest of the deck. It should also be noted that Air Canada has modernized its fleet, replacing its three-pilot aircraft with a two-pilot operation. Now a captain is in the left-hand seat, identified by four stripes on his epaulettes, and a first officer (or copilot), who has three stripes, is in the right-hand seat. Some long-haul flights employ one or two additional pilots to provide an opportunity for the operating captain and first officer to rest. On every flight the pilots are assigned one of two roles: PF (pilot-flying) and PNF (pilot-not-flying). The PF controls the aircraft while the PNF operates the radios and does the paperwork. These roles are usually reversed on the next flight, at the discretion of the captain.

Pilots like to incorporate familiar landmarks or picturesque sights in their announcements. On a flight from Boston to Toronto, they'll draw the passengers' attention to Niagara Falls. While navigating over Japan, the spectacular sight of Mount Fuji will be highlighted. To and from Europe, pilots will point out the vast snow-covered expanse of Greenland.

Routine sights are complemented by fiery sunsets or sunrises, the northern lights, many rainbow-like sundogs (parhelia), and halos around the sun or moon. A glory can often be seen surrounding an airplane's shadow. A glory is a pale, circular-looking rainbow caused by the scattering and reflecting of sunlight from water droplets in clouds below the plane. It's not a true rainbow and is usually very small, but it's something to look for during your next flight. If conditions are right, you can see the shadow of the airplane in this circular phenomenon as it dances along in unison with the airplane's path. Other spectacular sights include the phenomenal light show from a thunderstorm. Even when viewed safely from many miles away, the raw power of a thunderstorm elicits a deep respect for nature.

Occasionally, pilots are asked if they have ever seen anything strange in flight. One truly unusual sight was that of a California truck driver who tied 45 weather balloons to a lawn chair and soared to 16,000 feet (4877 m) in 1982. As he drifted near Los Angeles, this determined adventurer succeeded in disrupting air traffic in the area. While he took along a camera, he later admitted, "I was so amazed by the view I didn't even take one picture." The pilots who spotted the improvised aircraft undoubtedly shared this sense of amazement. When he finally landed his craft, the inventive pilot was the recipient of a substantial fine.

Stunning views are one of the many perks that come with a job that has its very own room with a view.

CHAPTER FIVE

Predescent
and In-Range Checklists

Landing at Airports Around the World

This flight has been routine. Flown with a complement of 14 flight attendants, all of them perform the first food service and then they break into two teams with each team having an on-duty service director (see page 60). Their crew-rest facilities are located below the main aft deck, and you may see them surface with company-issued pajamas with "crew" placarded on their backs.

On our navigation display, the marker denoting when it's time to start descending comes into view. Because of its shape on the Airbus, it's called the hockey stick. It's officially time to think about landing. But before we actually push the button to start the descent, an approach briefing must be given. The pilot-flying briefs the pilot-not-flying on which runway will be used, the type of instrument approach, flap setting, and lots more. The briefing tends to be generic, but it's done on every flight as everything in the flight deck is recorded. With the briefing out of the way, a predescent checklist is completed.

Predescent checklist

1. Approach briefing. This includes the following information: runway, type of instrument approach, flap setting, crossing altitudes at specific navigation points.
2. Ensure the navigation devices are loaded.
3. Pressurization check: ensure the automatic system is set properly.

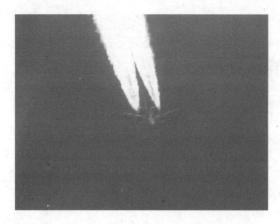

An aircraft passing 1000 feet above on the highway in the sky

4. Navigation accuracy is confirmed.
5. All systems working.

Flying for a large international airline means a pilot will inevitably fly to an airport they have yet to visit. Many airports come with their own inherent challenges. For some it's the weather; others may be nestled among mountains; for others it may be shear volume of traffic; and for others it may be a communication challenge due to the controllers' thick accents. However, by sticking to protocols, landing at a new airport is just as smooth as landing at your home airport.

Destination Unknown

There's a good chance that, when traveling by plane, you're on your way to a place you've never been to before. Though the prospect is exciting, you may have a few worrisome questions: Will the customs procedure go smoothly? Will the language barrier cause problems? And how will I find my way around the airport?

Pilots also land at airports unfamiliar to them but, luckily, protocols keep our anxiety to a minimum. For one thing, with few exceptions, English is the universal language for both pilots and air traffic controllers, a bonus for native English speakers. Standard terminology is used to ensure information is communicated clearly and safely.

Besides learning the industry terminology, an Air Canada pilot must go through a qualification process before he or she can fly overseas. The process involves reading briefing notes, listening to a lecture on survival equipment, and undergoing a computer-based training program, along with a simulator exercise made up of oceanic and diversion (emergency) procedures. Once completed, the training is followed by a route check in which a check pilot (supervisor) ensures that all the protocols have been understood and followed. Though the qualification process for domestic routes is slightly less rigorous, it still requires the pilot undergo a minimum number of supervised flying hours and a final flight with a check pilot.

In addition to this training, pilots flying to new destinations can log onto Air Canada's flight-operations website, where they will find information on local weather and runways, as well as any special procedures required for a particular airport. (Think of this website as a travel guide for pilots.) There is also an onboard library located in the flight deck. It contains information on international differences in aviation protocol.

Pilots also carry their own charts and maps for specific airports. (My flight bag keeps getting heavier and heavier as new destinations are added to my schedule, though a state-of-the-art electronic flight bag will soon replace it.) These visual aids depict the layouts of taxiways, runways, and gates, while taking such factors as noise abatement into consideration; at many airports, specific routes and altitudes are used to minimize the noise.

Throughout my career I've flown to more than 120 airports. Hopefully, you find landing at a new destination as exciting as I do.

The weather in Hong Kong is just as forecasted, with scattered thunderstorms lurking in the area, according to the Automatic Terminal Information System (ATIS). Available at most major airports in the world, this recorded aviation report broadcasts the latest weather, runways in use, and other pertinent information for pilots.

Looking outside the window confirms thunderstorms on the horizon. We are now at 40,000 feet (12,192 m), and by the looks of things the cloud tops appear higher. It shouldn't be a problem, though, as we can pick our way around them using our onboard weather radar.

The cardinal rule for real estate — location, location, location — also holds true for weather. One airport famous for inclement weather, due to its location, is St. John's, Newfoundland. It takes top prize for the rainiest, windiest, cloudiest, and foggiest airport in Canada. Many Canadian pilots have their dark-and-stormy night

On final approach to Runway 27 Right at Heathrow

story, and it usually originates from The Rock. So how do pilots find the runway in poor visibility?

Foggy Landings

Have you ever looked out of an airplane window as it descends, and you go lower and lower, and wonder when, and if, the ground will appear? Many of us have probably been on flights like this, but just how do pilots find the runway?

A Pilot's Approach

Despite what seems to be a precarious situation, commercial, and some private, pilots routinely fly safely into clouds with the aid of instruments. A handful of different instrument approaches are currently available, but the most precise and preferred approach is the Instrument Landing System (ILS), which provides both vertical and horizontal guidance in low-cloud conditions, fog, rain, snow, haze, and other obscuring phenomena.

How does it work? A localizer signal at the far end of the runway guides the pilot or autopilot in a straight line toward the runway, while a glide-slope signal on the sides of the runway leads the aircraft down vertically. An easy way to visualize a precision approach is to picture a children's slide at a park. The aircraft flies at altitude just as a child sits on top of the slide. The airplane is then steered in the direction of the runway, whereby the flight deck instruments lock on to both the localizer and glide-slope signals. When the aircraft is locked onto both signals, it is as if the airplane is in the crosshairs of a rifle. On board, sophisticated autopilots guide the aircraft all the way to the ground, automatically compensating for changing winds and other variables. The precision approach guides the pilot down to his or her landing sight (the runway), just as the slide guides the child to the landing. A localizer provides left-right orientation with the runway, similar to the sidewalls of the slide. The angle of this approach is typically three degrees. It's the angle you may have noticed airplanes maintain while following one another on approach to a busy runway. The airplane is literally dragged toward the runway by the two ILS signals on the ground. The glide-slope signal guides the aircraft down vertically, and the auto-thrust system adjusts engine-power settings to ensure proper speed, even bringing the engine to idle at touchdown.

Other Important Features of ILS

Several other components augment the ILS and provide additional safety features for low approaches. These include devices that transmit exact distances from the runway, high-intensity runway and approach lighting (the intensity ranges from a dim setting of one to power-zapping strength five), and radio-beacon markers that transmit important distances to the pilot. One such marker is called the final approach fix (FAF), which is typically located 4 to 6 miles (6.4–9.7 km) from the airport. At this point the pilot should have the landing gear down, a clearance to land from the control tower, and final flap settings for landing. Sitting by itself is a runway-visual-range (RVR) sensor along the edge of the runway, which alerts when visibility begins to deteriorate. It measures distance seen through obscuring weather phenomena in units of feet, and it gives a very accurate idea of what a pilot can expect to see, or not see.

Not All ILSs Are Created Equal

There are three different categories of ILS, differentiated by their decision height (DH) and prevailing visibility. DH is the indicated altitude at which a pilot must decide to either continue the approach to a landing or abort it. A category I ILS (the least accurate) has a DH of 200 feet (61 m) above ground. Most large airports around the world have this type of ILS. DH is determined by a barometric altimeter, which the pilot must adjust to the most recent pressure reading at the airport. Every pilot knows just one-tenth of a change in pressure in inches of mercury translates into a discrepancy of 100 feet (30.5 m).

A category II ILS has a lower decision height, 100 feet, and it determines height with a device that bounces signals from the airplane to the ground and back, called a radar altimeter (or radio altimeter). It allows the airplane to descend with a higher safety margin. The last, but certainly not the least, is the category III approach.

Welcome to Auto-land

Category III ILS (auto-land) has two levels. The first level brings the aircraft to a mere 50 feet (15.24 m) above the runway, at which time the pilot must make a snap decision. The second fully automated level has no decision height, meaning pilots do not look outside and wait for the bump. (It is a procedural necessity: pilots looking outside could cause them to become disoriented. Complete faith is bestowed in the system, which admittedly takes some getting used to.) A gamut of requirements must be met to allow such an approach. The ground facilities must have high-intensity runway lights, centerline lighting, various markings on the runway, additional RVR sensors, and backup airport emergency power to ensure the runways and taxiways are lit up and the ILS is functioning, even during power outages. On board the aircraft, sophisticated autopilots bring the aircraft to the ground, automatically correcting for winds all the way to the touchdown. Only major airports have such a system, with most only having the system on one runway. (Vancouver and Toronto Pearson have the only category III runways in Canada.) Pilots must be certified to do auto-lands, requiring checkouts in flight simulators every six months. The airline company and aircraft must also be certified for auto-lands. As you can see, there are a lot of parameters that must be met, clearly separating the amateurs from the pros.

For the airliner I fly, an auto-thrust system adjusts engine-power settings to ensure proper speed is obtained. In fact, it will even bring the engines to idle at touchdown. An auto-brake system supplies the correct amount of braking at touchdown to stop the aircraft. As well, there are many computers that monitor all of the aircraft systems to ensure everything is functioning at 100 percent. They even make synthesized altitude call-outs to the pilots.

Waiting for the Bump

The absolute minimum visibility for a category III landing is less than the length of a football field, with next to nothing to see when approaching at speeds of 150 knots (173 mph or 278 km/h). Once air traffic controllers clear the aircraft for a category III approach, the pilots attentively monitor the automatic systems, overpowering the urge to look outside, and patiently wait for the bump that indicates their safe landing. Even with the main landing gear firmly on the runway, the flight deck may still be mired in fog because of the landing angle. From ab initio training pilots are taught to trust their instruments; still, auto-land bestows a much higher level of faith in technology.

Because the system is so accurate, the automatic pilot must be disengaged after landing or else the aircraft will try to reposition itself back on the centerline of the runway. Finding the terminal building in such heavy fog can be a difficult task, but many airports have bright green lights embedded in the taxiways to guide the pilots to the gate.

The auto-land system truly is a marvel of technology and exemplifies just how technically advanced aircraft and airports have become. Nothing can replace the skill of an experienced pilot, but when extremely poor visibility dictates a category III auto-land, technology rules.

Seeing Through Foggy Windows

When it comes to the weather, all that most people care about is the temperature. But for a pilot, visibility is at the top of the list. When fog rolls in at the airport, a pilot's best friend is the runway-visual-range (RVR) sensor. This invaluable instrument helps to ensure safe takeoffs and landings by estimating how far down the runway a pilot can see. The current model was introduced in Canada in 1993, after three years of testing at the country's foggiest airport: St. John's, Newfoundland. The RVR sensor was a welcome replacement for transmissometers, the old instruments that measured visibility on runways across the country. Not only was the new sensor less expensive, but it was also more reliable and more accurate than the

Forward-scatter RVR sensor

old machinery, which required constant recalibration and sometimes stopped working due to mechanical problems.

Located off to the side about 1000 feet (305 m) from the end of the runway near the touchdown point, the RVR sensor sits on a 10-foot (3 m) high pole — the average height of a flight deck. The instrument, described by one engineer as a coat rack, is made up of two heads that emit light and two receivers that calculate how much of that light is scattered. The RVR sensor costs about Can$80,000 plus an extra Can$20,000 for installation. Toronto's Pearson airport has 13 of these expensive coat racks.

When you drive in thick fog and see the headlights of an oncoming car, you may also see a halo effect caused by the fog scattering the light in all directions. The thicker the fog, the greater the light diffusion. The same principle applies on the runway; the sensor is there to record this scatter effect and convert it into RVR. It then automatically relays this information to air traffic controllers and controllers in the radar center, as well as to Environment Canada.

The higher the scatter factor, the lower the visibility. When visibility is reduced to one mile (1.6 km) or less, the air traffic controllers broadcast the RVR to landing and departing aircraft. Depending on a number of factors, including the specific runway and aircraft model, a decision is made on whether or not to proceed with the takeoffs and landings.

So the next time you're taxiing for takeoff, keep an eye out for the RVR sensor, one of the pilot's most trusted allies.

Landing the Big One

Sunrise over the Atlantic

> Give briefing to other pilot on type of approach and runway to be used.
> Complete predescent and in-range checks.
> Follow instructions from air traffic control to fit in with other landing aircraft.
> Arm the approach button to guide the airplane vertically and horizontally down to touch point when cleared for approach.
> Slow aircraft by selecting desired speed and request first flap setting.

> Approximately 8 miles (13 km) out request flap two.
> Prior to the final approach fix, ask for landing gear to be selected down (i.e., lowered).
> Ask for flap three.
> Confirm the landing gear is down and locked by green lights.
> Keep autopilot engaged with auto-thrust engaged.
> Ensure landing clearance from control tower by hearing the words "cleared to land."
> Respond by saying, "Roger" when airplane's digitized voice tells you, "100 feet above."
> Airplane calls "minimums," the other pilot says, "runway in sight," and you respond by declaring "landing."
> Disengage autopilot.
> At 50 feet (15.24 m) above ground, the airplane tells you to retard the thrust levers.
> Simultaneously arrest the descent.
> Hold it off and wait for the bump.
> Allow nose gear to touch lightly.
> Engage thrust reversers; allow auto-brake to slow airplane.
> Disengage auto-brake, stow the thrust reversers, begin turning off runway with hand-controlled tiller.
> Wait for other pilot to say, "Nice landing."

An A340's landing gear

Smooth Landings

A few years ago, during a layover in London, England, I visited the Science Museum in South Kensington and gained a whole new appreciation for landing gear. Proudly displayed in the concourse near the main entrance is the right main assembly of an Airbus A340, the airplane I fly. Seeing the landing gear in this setting made me ponder its remarkable design, strength, and engineering. Put simply, this over 16-foot (5 m) steel-and-aluminum structure is a work of art.

The landing gear of a typical airliner consists of two main wheel assemblies, or mains, and one nose wheel in a tricycle arrangement. The B747, however, has four mains, capable of supporting one million pounds (453,592 kg). Each main consists of a large tubular strut, the wheels, and the brakes. Included with the strut is a network of rods and supports to sturdy it and to help tuck everything away when airborne.

The landing gear accounts for roughly 10 percent of an aircraft's weight. Hydraulic lines filled with fluid interweave throughout the innards of the airplane to move its massive struts. Supplying nearly 100 times the pressure found in a car tire, the hydraulic system retracts and lowers the gear. The engine-driven hydraulic pumps pressurize the hydraulic lines. The hydraulic system also operates the flaps, flight controls, and brakes.

The brakes — similar to some car brakes though much more

powerful — are found only on the main wheels. The nose wheel has no brakes but is the only wheel that pivots, allowing the pilot to maneuver on the ground. Found at the bottom of the main strut is a shiny oleo — a pressurized piston, which operates like a shock absorber in a car, to cushion the landings. It won't guarantee a smooth landing, but it certainly helps. A system like a car's ABS (anti-lock braking system) on the main landing gear prevents skidding.

Each movement of the landing gear, whether it is lowered or retracted, is called a cycle. Maintenance workers monitor these movements very closely, so they perform routine checks on the landing gear. During more rigorous maintenance checks, the airplane is hoisted on jacks that are strategically placed under it; this allows the landing gear to be cycled to ensure that everything is in top working order.

Generally only small pleasure airplanes have landing gear that doesn't retract. Airliners operate at much higher speeds, making it essential to move the landing gear away from the fast-moving slipstream. Just imagine the drag caused by large landing gear going at speed of 500 knots (575 mph or 926 km/h); fuel consumption would easily double. Putting your hand out of a window of a car traveling at high speed will quickly make you grasp the concept of drag.

Literally seconds after takeoff, the pilot-not-flying, or PNF, signals a positive climb by saying "positive rate," at which time the pilot-flying, or PF, orders the landing gear retraction. You may hear a noise after takeoff as the landing gear tucks itself into the belly or the root of the wings (for more on airplane noises see page 73). Now when you feel the airplane lifting off from the runway, perhaps you'll find yourself envisioning the pilots calling out "positive rate," followed by "gear up."

Activated by a lever in the flight deck with a small wheel attached to it, the landing gear is lowered about 3 to 6 miles (4.8–9.7 km) from landing. There's a lot going on to get an airplane configured to land, obtain landing clearance, and get it on the runway.

From basic training, pilots learn that three criteria should be met for landing: landing on the first one-third of the runway, landing in the center, and landing on the main wheels first. As one seasoned pilot said, "The rest is gravy."

Hong Kong air traffic control does a great job vectoring us around the thunderstorms, but we keep a close eye on things with our weather radar. Like everything in aviation, we perform our duties to standard operating procedures. This takes the guesswork out of the job. The aircraft is configured to land according to suggested distances, speed, and altitudes. (As one ornery boss of mine once said, "You can teach a monkey to fly as long as you have enough bananas.")

We touch down — the mains touch first and we feel a light thud, then the remainder of the bogie (the pivoting framework of the landing gear) makes contact, sending a firmer message that we have landed. I then slowly lower the nose wheel and arm the thrust reversers. It is decided full-reverse thrust will not be used today. The taxiway exits scoot by, and as the airplane slows the thrust levers are placed back into the idle position, the auto-brake is disengaged, and I turn the aircraft onto the taxiway by using a hand-controlled tiller, at which point the captain will take over and taxi to the gate.

The landing turns out to be a good one, but landings are like golf shots: sometimes you wish you had the chance to take it over.

"Sorry for the hard landing folks. It wasn't the pilot's fault, and it wasn't the plane's fault. It was the asphalt."

After a particularly rough landing during thunderstorms in Memphis, a flight attendant on a Northwest flight announced: "Please take care when opening the overhead compartments because, after a landing like that, sure as hell everything has shifted."

Another flight attendant's comment on a less-than-perfect landing: "We ask you to please remain seated as Captain Kangaroo bounces us to the terminal."

We receive a data link from crew scheduling half way around the world in Montreal. The scheduler is confirming the passport number for the fourth pilot on this flight. It's one of the toughest and more thankless jobs in the business ensuring all flights are crewed.

Crew Control

Air Canada flight crews hear a lot of things during their careers, but perhaps the most familiar words are "Hello, it's crew sked calling."

Every month, pilots bid for their schedules via computer. They are then awarded flights based on their seniority in the company and on the aircraft type (see page 2). Seniority comes into play for flight attendants, too. And don't forget the reserve crew, usually junior members, who must be available for contingencies like illness, mechanical or weather delays. These three groups make up close to 11,000 people, all of whom need to be slotted into flights — about 650 a day. That's where the crew scheduling group or "crew sked" comes in.

One day, I visited Air Canada's crew sked headquarters at Pierre Elliott Trudeau airport in Montreal. Duty manager Glenn Day introduced me to the eight schedulers in the newly renovated office. Large clocks hung high on the wall displayed the time in different zones around the world, while the computer screens and telephones worked overtime to keep Air Canada's entire fleet matched up with pilots. (Flight attendants have their own army of schedulers located in another building at the airport.)

Jason Quinnett was working at the desk that schedules the wide-bodied Airbus 340. He had little time to talk. He was fielding calls and had to find pilots to go to London, England. Though the list of potential pilots was dwindling fast, Jason kept his cool and eventually found someone to take the shift.

In order to find replacements, crew sked uses a variety of tools from the familiar — caller ID — to the specialized — a computerized scheduling system. The new programs used by Air Canada, CREWLINK and NETLINE, make up a modern and adaptable system. In addition to being wired in to communicate with pilots at all times, crew sked also has direct data link communication to each aircraft and the authority to ask telephone operators to cut into phone lines if necessary. As for the pilots, they are provided with toll-free numbers to link them up with the schedulers.

Before they can put out their feelers, crew sked members must know the rules inside out. Not only do they need to know all the Canadian Aviation Regulations (CARS) for scheduling, they also need to know about the latest pilot contract. For example, although there are many exceptions, a pilot generally can't fly more than 13 hours a day (in accordance with CARS) and can't exceed 85 flight hours per month (as per the pilot contract). What's more, schedulers must be firm yet understanding and show compassion toward ailing or grieving staff. They also act as travel agents — booking hotel rooms and arranging transportation. No wonder everyone agrees that crew scheduling is one of the toughest jobs at Air Canada.

This wide range of roles would also explain why the crew sked group has so many stories to tell. During my visit, one senior scheduler, Leo Lepore, recounted the story of how a pilot's wife called crew sked to say she was in labor and to ask if they could please locate her husband, a pilot laid-over in Fredericton, New Brunswick. He was not answering his phone. The front-desk staff at the hotel knocked at his room's door and the pilot was informed that his wife's water had broken. It took some doing, but the pilot made it home in time to be at his wife's side. When Leo related this tale to a more junior scheduler, the reply was "What do these pilots think we are? Plumbers?"

CHAPTER SIX

After-Landing
and Parking Checklists

Layovers, Traveling the World, and Beyond

Safely back on the ground, we are guided to the gate by the marshaler waving fluorescent wands. The park brake is set and all four engines are shut down. The cargo doors open, sending a signal with the official "in" time to operations. Today we traveled to the other side of the world and arrived one minute early. The Jetway moves into position, the passenger door opens, and salutations are exchanged between the ground staff and flight attendants. The passengers unload, a parking checklist is completed, and we pack up our gear for a well-deserved rest.

Many people, including my wife, think because aircrew fly all the time we don't get jet lag. As one senior captain said, "You always succumb to jet lag but you get accustomed to it." A coworker of mine claimed his neighbor would always challenge him to a game of squash the day after he arrived from overseas. He always lost. My motto is never operate anything dangerous like saws or lawnmowers the day after a long flight. Jet lag affects everyone differently, and knowing what works for you is the best remedy. For me, it's exercise. I've worked out in gyms all over the world: India, China, Korea, England. I remember working out in Japan immediately after a 14-hour flight, but that was pushing things; my lights were on but no one was home. If the gym sounds too energetic for you, just going for a walk will do wonders to clear your foggy brain. It's what most aircrew do, with many including mega shopping binges in their walks. (Yet another benefit of working for an airline: international shopping.)

Back on firm ground with the help of the B767's landing gear

The Pro's Guide to Jet Lag

You know the symptoms: lethargy, dehydration, disturbed sleep, and trouble concentrating. Doctors call the condition circadian dysrhythmia, but the rest of the world knows it as jet lag. It occurs when your built-in clock is out of sync with the outside environment, upsetting your bodily functions. Jet lag has been blamed for athletes losing competitions, politicians blundering international relations, and business people failing to secure major deals. Most people, including flight crews, succumb to it in some form or another.

Dr. Claude Thibeault, Air Canada's senior director of occupational health services, frequently advises Air Canada employees on how to cope with jet lag. According to Dr. Thibeault, there is no magical fix for everyone, but measures can be taken to minimize the effects of jet lag. And much of the solid advice he offers flight crews also applies to passengers.

Before the flight, Dr. Thibeault recommends plenty of rest. Many passengers think they can catch up on their sleep during the flight, but it doesn't always work out that way. For a night departure, consider taking a nap for one-and-a-half to two hours in mid-afternoon, which is a low period in your circadian rhythm. On the day of your flight, show up to the airport early, wear comfortable clothing and good walking shoes, and have everything you'll need, such as tickets and passports, readily at hand. Airports are getting larger and more congested, so they can present a real challenge to passengers who are running behind schedule, and a stressed traveler is not a rested traveler.

During the flight, get as comfortable as possible. Take off your tie, loosen your shoes or, better yet, bring slipper socks. One common strategy to minimize jet lag is to avoid alcohol and drink lots of water during the flight, as this reduces dehydration. Consuming water is so important that members of flight crews are given an extra liter (around 34 ounces) for every eight hours of duty.

Passengers can also follow the example of flight crews by stimulating their circulation during the flight. If you see a pilot walking down the aisle, don't be alarmed, he or she is probably just limbering up. At the very least, be sure to get out of your seat, if only to walk to the washroom. Some passengers want to stay awake during the flight to maintain their normal sleep schedule (or because they don't want to miss the meal service). "Studies have clearly demonstrated that the most successful technique for combating sleepiness is physical activity," says Dr. Thibeault. You can also keep alert by reading, playing a game, or conversing.

Once you arrive, your ideal sleep schedule depends on the length of your stay in the new time zone. If you are staying for only a day or so, as is the case with most crew layovers, it's better to just keep close to your home sleep schedule. But if you're staying longer you should heed the motto "When in Rome, do as the Romans do." Change your watch to the new time and try to adapt your sleeping habits accordingly.

Dr. Thibeault suggests that if you don't fall asleep in 20 to 30 minutes, you should get out of bed so as not to associate bed with sleeping problems. Get up and read or watch TV, but don't use alcohol as a sleeping aid, because it upsets your natural sleep patterns. Incidentally, you don't necessarily need one long sleep; two separate periods of deep sleep can be just as refreshing. The best way to ensure a sound sleep is to optimize your sleeping environment. Set a comfortable temperature, preferably on the cool side. Darken the room — wear a mask if necessary — and shut out any noise by using earplugs. Some crew members even refuse to take hotel rooms near the elevators and ice machines.

Working out can also help take the bite out of jet lag. Studies have shown that people who are fit tend to overcome jet lag faster than those who are not. Some members of flight crews make a point of heading to the gym during their layovers. It is a good idea, however, to avoid strenuous exercise immediately before sleep.

Everyone gets jet lag; it is a matter of personal difference as to how long you suffer after the flight. NASA suggests that it takes one day for each time zone crossed to recover from jet lag. The most effective approach to fight it may be to combine multiple strategies rather than to rely on a single one. Take your cue from flight crews, who are jet lag experts, to keep the traveler's enemy at bay.

And give up aviation?

I'll never forget how, during a downturn in aviation, we pilots met at a company meeting to hear how the company would be implementing layoffs. A somber mood dominated the group, but during a break a fellow pilot proceeded to tell me an anecdote that summarized the situation perfectly.

A pilot doing his routine walk-around neared the rear of the airplane where a ground worker (or lavologist) was servicing the lavatory. He was a low-time pilot trying to get his foot in the door by working the ramp, a common practice. As he released the latch to the aircraft's lavatory, the contents sprayed all over the ramp and unfortunately covered the wannabe pilot from head to toe. He cursed while jumping up and down in frustration and disgust. The pilot, witnessing this bathroom blowout, came right out and said, "Why don't you just quit?" The rampie stopped dead in his tracks with a look of bewilderment and said, "And give up aviation?"

Put Yourself in the Cockpit

Did you ever want to soar off into the distance and watch the realities of life fall away? An enthusiastic yes to that question makes you a good candidate to become a pilot, an immensely enjoyable pursuit that tests your limits, knows no boundaries, and gives you a brand-new perspective from above the clouds. There is often one small problem that confronts someone who wants to earn a pilot license: How do you get off the ground in the first place?

For those pining for the skies because they either want to be an airline pilot or just want to pilot their own bug-smashing two-seater for sightseeing, just get going. Your new career or hobby is just a phone

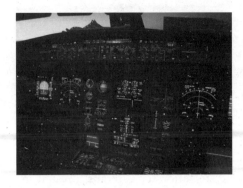

My "office" — the Airbus 330 flight deck

call away. To see if this endeavor is really for you, give your nearest flying club a call. All of them offer a familiarization flight. Some may find small planes a little cramped and the ride a little bumpy, making them realize they want to keep their feet well planted on terra firma, but I guarantee most will be wide-eyed, reassured that flying is for them.

The basic requirements for a pilot license are simple: You must be in good health and able to read. As long as you pass the medical exam, which validates your license, you can fly. For some the medical exam may reveal unwelcome surprises that clip your wings prematurely. One ailment afflicting some 10 percent of males is color blindness. Having said that, a friend of mine thought his aviation career was finished before it even started for that very reason, but further tests gave him the green light. He is now a jumbo jet captain flying in the Middle East. Many people think one must have 20/20 eyesight to be a pilot. Not so. I knew of two monocular pilots. It's true the military requires perfect eyesight at first, but again, getting a medical may open a door you thought was shut.

Mechanical aptitude also helps, but the will to learn far outweighs anything else. Obtaining a license entails a minimum of 45 hours of flying time: 17 hours in the company of an instructor, 12 hours solo, and the rest either by solo or dual flights. Any 14-year-old can fly solo in Canada. In the U.S., 16 is the minimum age for solo flights. In fact, some students earn their pilot license — age 17 minimum in both the U.S. and Canada — before they even get their driver's license! On the other hand, there are many pleasure pilots who are well into their 70s. (Pilots hired by Air Canada must possess the following: a valid Airline Transport Rating license and 3000 to 5000 hours of commercial experience. Having a degree, diploma from an aviation college, or military experience is a definite plus.)

The first thing you need to do is find a flying school and ask the students as many questions as you can think of. They will be able to tell you about the quality of the instruction, the availability of training aircraft, and what to expect from the school in general. Students can also help you choose an instructor who gives thorough briefings, both before and after flights. True aviation comes with a price tag, and in North America, the cost of obtaining your license is around $8000. Canadian students can now apply for student loans for most flight schools and colleges, and for those continuing on commercially, the cost is income-tax deductible. In the U.S., programs vary from state to state but there are many scholarships and bursaries available as well as subsidization for veterans through the Veterans Benefits Administration. Most good schools use the pay-as-you-fly method, which avoids hidden costs and additional hassles. Be wary of those that ask for your money up front or claim they can train you in a very short time. While it's true that you can obtain a license in seven weeks, you should plan to spend six months to a year completing the course — a calculation based on the availability of aircraft, the suitability of weather, and so on.

Once you've enrolled in a school, your instructor follows a standard published syllabus. Your first lesson will have you doing straight and level flights, followed by climbs, descents, and turns. After a few lessons you'll be able to make steep turns, stalls, spiral dives, and spins. Then, when you least expect it, your instructor will let you fly the aircraft by yourself, for your first solo flight — something no pilot ever forgets.

I remember my first solo. Talk about a confidence booster! (It's probably why many are christened with a bucket of water afterward to cool them off.) I can still see my instructor getting out alongside the runway and saying, "It's all yours" some 27 years ago. The feeling is probably similar to how chicks feel as they're tossed from the nest for the first time.

Tutoring in navigation, map reading, and instrument training will lead to your flight test, conducted by a government agency or by the school's own designated flight-test examiner. Most students easily pass because their instructor ensures that they only take the test when they are ready. Learning to fly also involves spending time on the ground, studying the finer points of navigation, air regulations, meteorology, flight theory, and engine mechanics. Courses are usually held at night and take about two months to complete, culminating in a 100-question written exam.

If getting a private pilot license is too expensive or too lengthy a process for you, there are recreational pilot permits. In Canada, the requirements are 20 hours of instruction and five hours soloing at a minimum cost of Can$4500. Some restrictions apply to this permit, however, which prevent you from flying outside Canada, carrying more than one passenger, or amassing the ratings that record your developing expertise. (Building up ratings allows you to move up to night flying, for example.) Most beginners opt to be certified as a private pilot, particularly if they want to eventually obtain a commercial license, which requires another two years of training at an additional cost. A commercial license allows you to actively pursue a career in flying, always a worthwhile occupation and a profitable one these days, with a five percent growth rate forecasted for the industry over the next 20 years.

Whether for fun or for a career, getting a private pilot license is well worth it. Imagine that the next time you hear "Cleared for takeoff," it may be for you!

You can obtain a pilot license in 7 weeks, but plan to spend 6 months to a year completing the course.

When you least expect it, your instructor will let you fly the aircraft by yourself for your first solo flight.

Getting your private pilot license is just the start. Next is the commercial license that makes it legal to fly an airplane for hire. This will not put you in the modern cockpit of an airliner making the big bucks, but it will allow you to fly for a small operation making poverty wages. I never said it was easy. The commercial license includes an endorsement to fly at night, but not to fly an airplane by instruments. You'll need an instrument endorsement to fly in clouds and inclement weather. Next you'll want to sweeten the pot by learning to fly a twin-engine airplane. With this you'll have a multi-engine instrument rating (the bare minimum requirement) and about 200 to 250 hours of flight experience. Now it's time to pound the pavement. At this stage many become flight instructors — requiring another endorsement and more money — to gain valuable flight experience. I did the instructor thing part-time, and eventually I had enough experience for a local cargo/charter company to hire me.

Luckily pilots are in big demand. Major airlines are hiring, which means bottom-feeder spots — a place you'll probably end up to gain valuable flight hours — are opening up. Once you have 2000 hours or so under your belt you can start applying to the heavy hitters, but don't expect to be called right away. Competition is stiff, but if you are willing to go abroad the world is your oyster. Places like China, the Middle East, India, and Europe are screaming for pilots. If there ever was a time to become a pilot, it is now. Here at Air Canada we have on average 10 pilots a month flying their last flight.

As mentioned earlier in this book, there are different ways to get spooled for an aviation career or hobby. Aviation colleges are great because they come with a diploma. The military is an excellent avenue to pursue; however, for some, the time commitment, not to mention other commitments that may be requested of them, makes this option less attractive.

If you're still in high school, another way of getting your private pilot license is through the air cadets. I taught many cadets to fly, and I also taught them meteorology — a two-for-one. What a deal for them. (From seeing the program firsthand, I wish I had taken the challenge when I was their age, but having short hair back then was not cool.) They get a private pilot license during the summer, compliments of the government, and many of them are still too young to have a car license.

Extreme Flying

For most pilots I know, the love of flying runs deep. But airline pilots aren't the only ones who rule the skies. What people outside the industry often don't realize is the diversity of professional paths available to aviation enthusiasts. Aside from airline pilots, there are many other kinds of pilots, including those who take flying to the absolute extreme.

Often celebrated and even mythologized, test pilots push the limits of air travel. It's a job that requires patience, curiosity, and nerves of steel, as well as the utmost piloting skill. Canadian Chris Hadfield displayed such skill when he received his test-pilot training at Edwards Air Force Base in California in the late 1980s.

Hadfield has won the two highest test-pilot honors in the U.S. military. Incidentally, the base where he trained also saw a remarkable moment in aviation history some four decades earlier, when test pilot Chuck Yeager first broke the sound barrier, as depicted in the movie *The Right Stuff*. Like many of the test pilots depicted in the movie, Hadfield went on to become an astronaut. In fact, he was the first Canadian to walk in space and remains one of Canada's most famous astronauts. (Incidentally, Chris has two brothers flying for Air Canada.)

Training astronauts like Hadfield requires yet another kind of pilot. To experience weightlessness without leaving Earth, astronauts-in-training ride in the back of a large KC-135 airplane while the pilot performs maneuvers that simulate zero gravity. The aircraft is flown to a high altitude and then sent into a ballistic parabolic dive, which the pilot pulls out of at a lower altitude. During the dive, those riding in the airplane free fall, temporarily experiencing a feeling of weightlessness. This aircraft is nicknamed the "vomit comet" for good reason. The minute or so of zero-G is usually used for astronaut training and equipment testing, but it was also used during the filming of Ron Howard's *Apollo 13*, for the movie's zero-G sequences. The crew filmed in one-minute segments inside the aircraft as the pilot flew parabolic arcs.

Another kind of pilot who pushes the limits is the hurricane hunter. In the name of science, these pilots fly into the center of some of the Earth's most destructive meteorological forces. The more that's known about how strong a hurricane is, how fast it's going, and where it will hit land, the more people can be protected. Hurricane hunters endure a very turbulent ride in order to measure the storm's size and strength. They fly C-130 Hercules aircraft that are adapted for weather reconnaissance and equipped with computerized meteorological data-gathering instruments. An onboard flight meteorologist observes and records data at flight level every 30 seconds. Vertical meteorological data are collected using a parachute-borne sensor, called a dropsonde, which measures and encodes weather data down to the ocean surface. Once the aircraft is inside the eye of the storm, calm flight conditions prevail and the sun even comes out — a sight very few ever witness. The only catch is that the airplane must eventually make the treacherous journey back outside. (See page 111 for more on hurricanes.)

The list of pilots doesn't end there. There are also jet fighter, glider, helicopter, ultralight, balloon, float plane, ski-plane, and aerial photography pilots, as well as cargo, medevac (medical evacuation), news filming, police, offshore surveillance, search and rescue, fire patrol, crop dusting, stunt, and private-business jet pilots. With so many kinds of pilots, the world of aviation is indeed diverse. Airline pilots are just one chapter in the handbook on pilots.

The flight deck of the future will require one pilot and one dog. The pilot will be there to feed the dog, and the dog will be there to bite the pilot if he touches the controls!

Maybe flying an airplane is not for you. Maybe just getting on an airplane is a challenge. Hopefully this book enlightened you on the intricacies of an airplane and the business of flying, but what if it's still not enough to calm your nerves before your next flight? That's okay, you are not alone and there are lots of options. One Air Canada pilot I know has taken the initiative to offer fear-of-flying courses — Captain Marc-Antoine Plourde started Deplour Research and Training Centre (www.deplour.com) — and there are many more programs available across North America.

Have No Fear

Nearly 30 percent (some claim 40 percent) of the adult population suffers from aerophobia, or aviophobia, the fear of flying. It has also been estimated that one in four flyers shows a significant degree of fear or anxiety. And it's not only the less traveled who fall into this group. Many executives and celebrities who rack up thousands of air miles are secretly afraid of flying. Surveys indicate that most fearful flyers do fly and many are frequent flyers. Why the fuss, since no other form of travel — not even walking — approaches the modern jetliner in safety?

It only takes a few minutes of surfing the Internet to realize that an abundance of information and courses are available. There are tapes, videos, newsletters, groups, and so on. Most major airlines offer a course and Air Canada is no exception. Their tactic is to strategically overcome the fear of flying by knowledge and control. With the contribution of psychologists, the instructors have a very high success rate, with graduation consisting of a flight on board Air Canada.

For many not being in control is a major hurdle. Learning to overcome fear of flying is really about learning to overcome fear of being a passenger. When we pilots travel as passengers, many ask if we mind sitting in the back where we are not in control of the plane. To answer that question, ask a flight attendant; they'll tell you pilots are not always ideal passengers.

Knowledge is power. The classroom part of the course for fearful fliers is equally divided between aviation education and behavior modification. You will certainly end up knowing more than the average passenger, with topics covered including the theory of flight, meteorology, and a particular focus on turbulence, since it is a frequent cause of anxiety. Although turbulence can cause discomfort, it is important to remember it is not unsafe when you are buckled up (see page 82 for more on turbulence). An explanation of the aircraft components will enlighten the student on the different noises aircraft make so they can learn to recognize the sound sequences (see pages 73 for more on aircraft sounds).

Still not convinced? When asked whether we pilots sometimes find flying dangerous, we respond by saying "The most dangerous thing about flying is driving to the airport." My three fender benders in taxis and crew buses make me a firm believer. If you are craving more statistics, well it's 100 times safer to fly than to drive; you are more likely to die by being kicked to death by a donkey than in an air crash; and a fully laden jumbo jet would have to crash every day with no survivors to equal the number of road deaths in a year in the United States.

A common question is how many hours the flight crew typically gets on layovers. If the flight is daily, it translates into about 24 hours. Sometimes it works out to be two, three, or even four days, depending on the schedule. The time of year dictates whether a daily flight is necessary or if two to three times a week will meet passenger demand. This applies to international layovers. A domestic flight may mean a much shorter layover, occurring at the airport hotel, which doesn't go down too well with some.

Another question high on the priority list is what transpires on layovers. I've worked for some upstart airlines, and many of my fellow pilots were young, with aviation being their first major job. And, yes, there was certainly some dipping of pens in company ink. I've seen many relationships develop into marriages and unfortunately I've also seen some falter.

With most large airlines the back-end and the front-end staff (see page 59) have been around for many years, so they just show up for work and mingling becomes more and more rare. Many turn into slam clickers, aviation lingo meaning that on hotel layovers they "slam" the door and "click" the locks. You never see them until crew pickup.

In one particular layover town, hundreds of airline crew congregate every night, which requires many hotels to house them on layovers. Because of the numerous time zones crossed by most and the resultant irregular sleeping patterns, not to mention the odd drink being imbibed, the town has seen its fair share of aircrew letting off a little steam. In order to contain the raucous, the locals developed a meeting place that consists of a tractor-trailer with bar amenities where crews have been known to party to the wee hours of the night. I had been to this town over 30 times but never paid this infamous landmark a visit . . . until this Christmas Eve. In the name of investigative journalism and feeling in the festive mood, I finally went and now I can say firsthand the visit was worth it.

Most jobs have their perks, and for me, as an international airline pilot, visiting neat places around the world is one of them. Swimming in the Dead Sea, taking day trips on London's Tube, visiting the largest wooden temple in Japan, and shopping in Hong Kong have all proven to be memorable. But one such unforgettable and glad-to-be-alive-to-tell-about-it trip was the bus drive to the Taj Mahal.

Trek to the Taj

Our 10-hour flight from London, England, landed in New Delhi at midnight. The crew layover was two days, but making the tour to the Taj Mahal required getting up in the early morning hours on the first day. The bus departed the hotel at 6:15 a.m., a mere five hours after I had settled in at the hotel. The 125-mile (200 km) trek to Agra would take just under five hours. Only 10 people from various hotels boarded the small bus, and they all appeared to be airline types. Just one flight attendant from my crew thought it was worth the sleep deprivation to see this great wonder of the world.

The ride started off somewhat passively as we watched the city come alive, although it did include the sight of people squatting or standing along the roadside to do their morning ritual. It was impossible to catch up on sleep because of the rough ride. It wasn't just that the roads were in terrible shape, the bus was only limited to about 45 miles per hour (70 km/h), and at this top speed it nearly reverberated to self-destruction. The horn was used continually. In fact, most of the trucks, of which there were many, had signs on the back stating, "Horn please!"

There were two mandatory stops to pay road taxes, and both times the bus was approached by peddlers, snake charmers, or beggars wanting money to take pictures with their black bears (the bears don't wear muzzles) or their monkeys. The ride was interesting to say the least, with numerous overturned trucks looking like turtles on their backs never to be uprighted. Most of the trucks had large wheels, which inherently causes the center of gravity to be significantly higher aboveground, making them susceptible to overturning. There were burned-out buses (I hoped it wasn't an omen), and the rotting carcasses of camels, goats, and other unrecognizable animals. Much of the road was divided, but oncoming traffic still ventured into our lane making things even more interesting. The highway was littered with ongoing construction, making for very tight maneuvering around gaping holes and equipment. Gas stops were required, but every place we went to had local power outages, so the fuel had to be pumped manually.

Finally we arrive! As soon as we disembarked we were immediately bombarded with street vendors offering everything from T-shirts to pictures of you with the Taj in the background. The day we traveled, Friday, was a popular day for visitors and lineups quickly built up.

The Taj Mahal, with its vastness of white marble, is a thing of beauty. Sunglasses are highly recommended because of the sun's strong reflection off the marble. Because of the continual threat of "hawkers," you may find it hard to relax and enjoy its splendor. We were told valuables could be left onboard the bus, but that sugges-

Safely arrived at the Taj Mahal

tion was met with skepticism.

After touring a few other locations in the area, we clambered back on the bus for the ride back, which proved to be even more interesting as most of it was in the dark. Cooling at dusk allows the hovering layer of smoke and pollution to blanket the land, significantly reducing visibility. Many vehicles had no headlights, few had brake lights, and turn indicators were unheard of; no wonder it is considered one of the world's most dangerous roads to travel. I have been in very strong turbulence, landed in near-zero visibility in blinding snowstorms, taken off in viciously gusty crosswinds, and I can say with authority that this drive is not for the faint of heart. Again, the driver honked the horn for the four-and-a-half hour return. He must be commended because he had to be on his game the entire time. The bus finally arrived at its last stop, our hotel, at 11:15 p.m., exactly 17 hours later.

The Taj Mahal truly is one of the wonders of the world, and surviving the drive was a wonder in itself. I would certainly see it again, but next time I'll take the train.

✈ ✈ ✈

As discussed, a remedy for jet lag is exercise, even just getting out and walking. During a recent layover in Vancouver, I decided to capitalize on the rare lengthy stay and check out the raved-about Grouse Mountain, which overlooks the city. By climbing and touring Grouse Mountain, I helped rid myself of jet lag by getting a great workout and touring the area — two remedies in one.

Grueling Grouse Grind

I've climbed Signal Hill in St. John's, Newfoundland (contending with strong winds in the windiest city in Canada); Mount Masada, near the Dead Sea in Israel (on the Snake Path one ascends from 900 feet/274 m below sea level to 200 feet/61 m above); Victoria Peak, Hong Kong (with the early morning temperature hovering near 86°F/30°C and the humidex off the charts). So I thought the climb on Grouse Mountain would be a similar challenge. Wrong!

With over a day's stay in Vancouver I decided to partake in the rage that has been sweeping Vancouver: the Grouse Grind, a.k.a. the Gross Grind! Once at Grouse Mountain a large sign greets you with a warning: people with health ailments should think twice about venturing up the hill, and plan about one-and-a-half hours for the ascent of 2800 vertical feet (853 m). I have just a hint of hypertension, but I'm also a regular in the gym, so the sign didn't intimidate me. I made sure I noted my time at the official start point, my heart beating fast in anticipation of climbing one-and-a-half times the height of the CN Tower.

The climb started out okay. The trail was well marked on the trees, but I soon realized that it would be easy to stray as I entered a hypoxic state, gasping for air. I missed the halfway-point sign and was pleasantly surprised to see the three-quarters-of-the-way sign, but not before thinking about turning back, and also wondering how they lift exhausted climbers or cardiac-arrest victims out by helicopter.

I passed people who were also gasping for air, stuck to the hillside like stunned mountain goats, and started to feel better about myself. Believe it or not, there are even people who take this torturous path back down, but it's certainly not advisable if you are concerned about your knees. There are no flat parts, which would have allowed your heart rate to slow down, and this can work on you both physically and psychologically. I met a few climbers along the way who were moving at my speed, and it was easy to start a conversation regarding personal times for the Grind. As I neared the top, soaked from head to toe, sunshine poked into the path — the proverbial light at the end of the tunnel.

For me the climb took 55 minutes, a climb a fully loaded jumbo jet could do in less than a minute. There's no doubt about it, the climb is exhilarating and you get that euphoric, glad-to-be-alive feeling. Make no mistake about it: the Grouse Grind is not for everyone. If you could do an hour on a stepper machine then go for it, but if you don't know what a Stairmaster is, then take the gondola — the view on top is worth it!

Greenwich, Zulu, and Universal coordinated time are all familiar ways to discuss the systems of time every pilot utilizes. As luck would have it, fine weather in London, England, a few summers ago allowed me to discover where time starts and the origin of worldly time.

The Beginning of Time

For me, as an overseas pilot, the question "what time is it?" is not always easy to answer. As I cross time zone after time zone, this simple question becomes more complex. On a London layover, my flying partner and I decided to visit Greenwich, where time originates every day, and where zero degrees longitude (the prime meridian) divides the Earth on its vertical axis. It's the exact spot where east meets west.

Greenwich Observatory

Getting there is easy. Greenwich is slightly to the east of London's city center, and it takes roughly 15 minutes to get there by Tube (London's subway system). The Royal Observatory is about a 15-minute walk from Greenwich station, through the town of Greenwich. There you'll see people snapping tons of pictures in front of the Shepherd Gate Clock, which depicts official time, and where you can get a great photo of yourself straddling the eastern and western hemispheres.

Precisely 1:00 p.m. at Flamstead House, Greenwich, England

Plan to get there before one o'clock local time so you can see the conspicuously painted red ball drop. This time ball has dropped every day precisely at 1:00 p.m. since 1833. The observatory had set a large time ball on the roof of Flamstead House to serve as a visual cue so navigators on the Thames could synchronize their clocks. Nowadays, overhead satellites send continual signals to update flight deck clocks. Why not drop the ball precisely at noon? The astronomers chose one o'clock because at noon they were busy with their astronomical duties, measuring the sun as it passed the local meridian. The ball rises halfway to the top at 12:55, reaches the top at 12:58, and drops exactly at 1:00 p.m., but without noise so be careful — blink and you may miss it.

What Time Is It?

Up until the mid-19th century, every individual town around the world kept only local time. In October 1884, it was decided Greenwich would be where time starts. Pilots, air traffic controllers, weather personnel, scientists, and mariners all converse in GMT (Greenwich mean time). The U.S. military named each time zone after the phonetic alphabet, and Greenwich was awarded Z for Zulu. Hence all users of standard time also refer to it as Zulu time, and you may see a Z affixed to Greenwich time.

To make things more complicated, GMT was replaced by Universal Time Coordinated (UTC) in 1972. (UTC is also referred to as Universal Coordinated Time or Coordinated Universal Time.) Thus world time is no longer set at Greenwich, but is now kept by time laboratories around the world using highly precise atomic clocks. Using these instruments replaces using the Earth's rotation, which inherently has very small time hiccups. However, keep in mind that the baseline of the international time zone system, and the place "where east meets west," is still in Greenwich.

UTC uses the 24-hour, or military, clock. For example, 3:00 p.m. in New York is 15:00 in military time. Now, if I add five hours (to account for eastern standard time) to 15:00 it is 20:00Z (UTC), the time in Greenwich. One would add just four hours during eastern daylight time. Confused?

Most do not realize the complexity of time. When I flew from Toronto to New Delhi, India, direct (a 14-hour flight), it was suggested we make passenger announcements near the beginning of the flight that reference time (we cross 11 time zones); this would ensure the local time given is correct, especially where India works on the half hour.

So, the next time someone asks you the time, ask them if they want it in local, GMT, UTC, or Zulu.

Another fun thing about being a pilot used to be flight deck visitations. Unfortunately, following the tragic events of September 11, 2001, new rules forbid visitations; a locked door has replaced the open-flight-deck policy. A very expensive bulletproof Kevlar armored door guards the flight deck. I used to enjoy visits. It broke the monotony and it gave us a chance to show off the flight deck.

Nowadays, passengers seem to be more in the spring-loaded position and rightly so. They tend to be more conscious of other passengers and their environment. Before, on long-haul flights I would walk the cabin to stretch my legs. Now when I do, I get worried and concerned looks and tend to cut my walk short.

Even flight attendants' visits to the flight deck are less frequent, almost non-existent. The first time we get to meet many of our crew is when we board the bus on the other side. The question

"Were you on my flight?" is popping up more and more often.

Some pilots claim much of the fun is gone from the job. Being isolated in the flight deck for hours, new procedures at security, being searched in front of passengers, restrictions on what you can carry in your flight bag — it all contributes to making a pilot's working environment less enjoyable. However, it's an aspect of aviation that will be a part of the job description for the foreseeable future.

Like the death of Elvis Presley or the assassination of JFK, everyone remembers where they were and what they were doing on September 11, 2001. Here's my story.

A Day of Infamy

The words "A Day of Infamy" shot out from the front page of the *Globe and Mail* to depict the horrific events of September 11, 2001. Coined by the media as "the Attack on America," it will always remain a dark day in history.

Halfway over the Atlantic Ocean, while flying from Frankfurt to Toronto, we began hearing escalating chatter over the air-to-air frequency of 123.45 (a frequency pilots use when flying over the Atlantic and Pacific oceans to communicate with other airplanes). Two days before, the topic of conversation was rather mundane and pertained to the turbulence encountered over the mid-Atlantic heading eastbound. However, on September 11th, the topic of conversation seemed surreal. Could I believe what I was hearing? A plane had crashed into the World Trade Center, and possibly a second plane crashed into it 15 minutes later? More and more conversations verified the information as data links from dispatchers began pouring into flight decks.

We listened to George W. Bush's short speech on BBC via high-frequency (HF) radio, acknowledging the attacks as the acts of terrorists. Ironically, the night of my layover I had been watching CNN at 3:00 a.m. Frankfurt time because of overseas insomnia, and they ran a documentary on the terrorist life of Osama bin Laden.

Nearing landfall, we learned that American airspace was closed. We data linked a message to our dispatch asking to confirm that the events were true, as it still didn't seem to be sinking in. It was verified and followed by instructions to avoid American airspace and lock the flight deck door. The service director was briefed, followed by the flight attendants. It was decided not to tell the passengers as we feared mayhem might break out. As westbound flights were identified by Gander radar they were told they had to land in eastern Canada. We could see airplanes making abrupt turns to places like St. John's, Stephenville, and Gander in Newfoundland, and Halifax, Nova Scotia.

While en route the captain decided to move the fire ax closer to make it readily accessible. I'm not sure if he meant to use it or to hide it from a potential intruder. He asked me if I could or would use it if need be. I said probably not, although one never knows what one would do when backed into a corner.

Our flight was allowed to continue, although we were re-routed. While over the Maritimes we found out the airports in Toronto and Ottawa, Ontario, were closed. We were told to land at Montreal's Dorval Airport (now Pierre Elliott Trudeau), but that had wavered between Montreal's other airport, Mirabel, Quebec City, and Bagotville, Quebec. Well into the descent, the captain made an announcement to the effect that we were landing in Montreal because the Toronto airport was closed, giving no details as to why. It's a known fact passengers want to hear the truth when faced with delays, diversions, or cancellations, but we kept it short.

Over eastern Canada the airways were eerily void of chatter. We discovered Air Canada had shut down operations completely, which until that time only a strike had accomplished. On descent into Montreal it was ghostly quiet with only a KLM flight sharing the frequency.

During taxi another very short announcement was made to let people know customer agents would be meeting them to answer all their questions. Rumors and wild stories flew as the doors opened and ground crew rehashed the day's events. Flight attendants were visibly upset, and tension was high among the passengers in the terminal. Flight operations found us a truck-stop-type hotel some 40 minutes from Montreal, and we considered ourselves lucky as stranded passengers inundated hotels. (A day later one pilot gave up and hitched a ride back to Toronto with a trucker.) Some 45 airplanes had landed in Halifax and the airport authority closed one runway to accommodate them all. In Gander, there were more people at the airport than residents in the town itself.

While watching television in the hotel room, graphic pictures of the World Trade Center crashing to the ground confirmed the atrocities of that day. Memories of my many layovers at the Marriott hotel based at the foot of the towers came back, and how I worked out in the hotel gym overlooking the Statue of Liberty. What was she thinking that day, seeing the foundation of peace and liberty crumble before her eyes?

That horrific day has made many realize that terrorism is upon us. It has made pilots realize they can become human missiles — something that, until then, was unthinkable when flying the very safe airspace of North America. As George W. Bush put it a day later, "Yesterday has changed tomorrow." It was a sad day for humankind and a sad day for aviation. That day of infamy will never be forgotten.

September 11th was unequivocally the biggest blow to aviation, and even today we are feeling and dealing with the aftershocks. Airlines have fallen to the wayside, with most airlines in the United States still not entirely stable. Many aviation experts feel one or more of the American majors must go in order for the others to survive. For everyone's sake I hope this will not be the case and they can pull it off together.

Prepare for Blastoff

Modern aviation began on December 17, 1903, when Orville and Wilbur Wright demonstrated to the world the first manned, powered, sustained, and controlled flight by a heavier-than-air aircraft. Although it lasted a mere 12 seconds and traveled only 121 feet (37 m), those few seconds changed the world.

The modern, fuel-efficient, fly-by-wire aircraft is a distant relative of the noisy, gas-hungry, piston-driven planes most of today's pilots flew at the beginning of their careers, and it's several generations removed from Orville Wright's machine.

The aviation business continues to fly high. Annual traffic growth of five percent is forecast for the next 20 years, during which time the major world airlines will more than double their current number of passenger seats. Two billion passengers a year are expected to fly by the end of this decade. This increased demand has brought on the development of mega-jets: next-generation super-jumbos that will have room for 800 or more people.

The technology already exists for these Goliaths, and Airbus has unveiled its new jumbo A380, which carries up to 650 passengers. However, the sheer size of mega-jets makes maneuvering them on the ground problematic at most airports; but many airports, including Toronto's Pearson, claim they are ready for them. Not to be left behind, Boeing has been modifying its 747, to stiffen the competition. The B747-8 is scheduled to be on the ramp in 2009.

Work also continues on supersonic transports. It is hard to believe the Concorde first flew in 1969, some 38 years ago; but then again, revenue passenger flights only started in 1976. The Concorde traveled at twice the speed of sound and was literally faster than a speeding bullet. Its nose temperature could go from 160°C to −10°C (320°F to 14°F) in minutes due to surface friction, and its body stretched 9 inches (23 cm) during cruising altitude. A consortium of plane manufacturers has produced plans for the so-called son of Concorde, although the problem of noise pollution caused by its inherent sonic boom will have to be resolved.

Beyond supersonic transport lies a new kind of plane, one that combines flight and space technology to reach speeds of more than 18 times the speed of sound. NASA's hyper X program, in collaboration with Boeing and others, built the X-43A, which busted the then-world-record for speed, Mach 10 (about 5000 mph). The plane was powered with technologically advanced engines called scramjets. Unfortunately the hyper X program faltered in 2004, but . Boeing and Pratt & Whitney recently began pursuing a new prototype, the X-51 Scramjet Waverider.

Plans are already underway for permanent space stations that will eventually house hundreds, perhaps thousands, of people. Manned outposts on the moon and even on Mars are also envisioned for early in the next century. Sound far-fetched? The history of flight is the story of the impossible turned into reality. No doubt the future will bring more amazing aerospace achievements. Don't be surprised if one day the ever-so-common phrase heard by pilots, "cleared for takeoff," is replaced by "cleared for blastoff!"